Proceedings of the Society of Photo-Optical Instrumentation Engineers

Volume 75

Imaging Through the Atmosphere

March 22-23, 1976 · Reston, Virginia

James C. Wyant
Editor

Presented By
The Society of Photo-Optical Instrumentation Engineers
and the Society of Photographic Scientists and Engineers

In Cooperation With
Optical Sciences Center, University of Arizona
National Oceanic & Atmospheric Administration, U. S. Department of Commerce
U. S. Army Electronics Command, Night Vision Laboratory
Energy Research & Development Administration
Institute of Optics, University of Rochester
National Aeronautics & Space Administration, Ames Research Center
Department of the Navy, Office of Naval Research

ISBN 0-89252-102-3

IMAGING THROUGH THE ATMOSPHERE

Volume 75

Contents

Seminar Committee v
Introduction vi

SESSION 1. Atmospheric Effects 1

A Review of the Optical Effects of the Clear Turbulent Atmosphere 2
Robert S. Lawrence, National Oceanic and Atmospheric Administration

An Elementary Derivation of Phase Fluctuations of an Optical Wave in the Atmosphere 9
H. T. Yura, The Aerospace Corporation

The Separation of the Optical Transfer Function in a Turbulent Medium 16
Wayne W. Metheny, Richard B. Philbrick, Optical Sciences Center, University of Arizona

Varieties of Isoplanatism 20
David L. Fried, Optical Science Consultants

Characterization of Atmospheric Turbulence 30
M. Miller, P. Zieske, Avco Everett Research Laboratory; D. Hanson, Griffiss Air Force Base

Dynamic Atmospheric Turbulence Corrections 39
Robert J. Noll, The Perkin-Elmer Corporation

SESSION 2. Atmospheric Measurements 43

An Experiment for Measuring Effect of Atmospheric Turbulence on a Vertical Optical Path 44
Robert R. Shannon, W. Scott Smith, Optical Sciences Center, University of Arizona

Stellar-Scintillation Measurement of Vertical Profile of Refractive-Index Turbulence in the Atmosphere . . . 48
G. R. Ochs, R. S. Lawrence, T. Wang, NOAA; P. Zieske, Avco Everett Research Laboratory

A New Turbulence Sensor Using Atmospheric Dispersion 55
Richard H. Hudgin, Itek Corporation

Turbulence Effects upon Laser Propagation in the Marine Boundary Layer 62
Kenneth L. Davidson, Thomas M. Houlihan, Naval Postgraduate School

SESSION 3. Speckle Interferometry 69

How to Build a Speckle Interferometer 70
Arthur M. Schneiderman, Douglas P. Karo, Avco Everett Research Laboratory

Astronomical Speckle Imaging 83
P. Nisenson, D. C. Ehn, R. V. Stachnik, Itek Corporation, Optical Systems Division

⇒

SESSION 4. Pre-Detection Compensation 89

Active Image Restoration with a Flexible Mirror 90
A. Buffington, F. S. Crawford, R. A. Muller, A. J. Schwemin, R. G. Smits, University of California

Using Membrane Mirrors in Adaptive Optics 97
Martin Yellin, The Perkin-Elmer Corporation

Wideband Adaptive Optics for Imaging 103
Julius Feinleib, J. W. Hardy, Optical Systems Division, Itek Corporation

Optical Wavefront Correction in Real Time 109
Virendra N. Mahajan, The Charles Stark Draper Laboratory, Inc.

The Effects of Atmospheric Dispersion on Compensated Imaging 119
Edward P. Wallner, Itek Corporation

Adaptive Optics for Space Telescopes 126
T. R. O'Meara, C. J. Swigert, W. P. Brown, Hughes Research Laboratories

SESSION 5. Post-Detection Compensation 135

Post-Processing of Imagery from Active Optics—Some Pitfalls 136
Richard E. Wagner, Optical Sciences Center, University of Arizona

Fundamental Limitations in Linear Invariant Restoration of Atmospherically Degraded Images 141
J. W. Goodman, J. F. Belsher, Stanford University

Digital Image Processing of Simulated Turbulence and Photon Noise Degraded Images of Extended Objects 155
James R. Breedlove, Jr., Los Alamos Scientific Laboratory

A Statistical Method for Post-Detection Compensation for Atmospheric Distortions of Images of Faint Scenes 163
Charles E. KenKnight, University of Arizona

Author Index 169
Subject Index 169

Seminar Committee

IMAGING THROUGH THE ATMOSPHERE

General Chairman
James C. Wyant
Optical Sciences Center, University of Arizona

Co-Chairman
Raymond P. Urtz, Jr.
Rome Air Development Center, Griffiss Air Force Base

Chairman Session 1—Atmospheric Effects
James C. Wyant
Optical Sciences Center, University of Arizona

Chairman Session 2—Atmospheric Measurements
J. W. Goodman
Dept. of Electrical Engineering, Stanford University

Chairman Session 3—Speckle Interferometry
R. Wagner
Optical Sciences Center, University of Arizona

Chairman Session 4—Pre-Detection Compensation
Raymond P. Urtz, Jr.
Rome Air Development Center, Griffiss Air Force Base

Chairman Session 5—Post-Detection Compensation
D. L. Fried
Optical Science Consultants

Introduction

Developments have been made in understanding the effects of the atmosphere on optical imaging systems and in eliminating, or at least reducing, these atmospheric effects on an optical system's performance. This conference emphasized developments in the understanding and measurement of atmospheric effects, speckle interferometry, post-detection compensation techniques, and adaptive optics.

Session 1
ATMOSPHERIC EFFECTS

Session Chairman
James C. Wyant
Optical Sciences Center, University of Arizona

A REVIEW OF THE OPTICAL EFFECTS OF THE CLEAR TURBULENT ATMOSPHERE

Robert S. Lawrence
National Oceanic and Atmospheric Administration
Environmental Research Laboratories
Boulder, Colorado 80302

Abstract

The refractive irregularities that are responsible for the atmospheric degradation of optical images arise from variations in temperature of the atmosphere. We present a non-mathematical, physically based description of these irregularities and how they contribute to the degradation of an image. Irregularities near the image contribute only through the distortion of the wave front, thereby distorting or spreading the image of a point source. More distant irregularities produce, in addition, scintillations or fluctuations in the intensity of the light. The random apodization resulting from scintillations further degrades the image.

Introduction

Refractive-index variations in the clear atmosphere affect light waves passing through them and, in most cases, degrade the performance of optical systems. We shall describe the refractive-index variations with particular attention to the small-scale fluctuations associated with atmospheric turbulence. Then we shall consider the effects of the fluctuations on optical systems, emphasizing the motion, blur, and scintillation produced in the image of a point source. Although we shall usually describe the effects in the context of the image of a star seen with a ground-based telescope, that description applies with only minor and obvious modifications to other situations such as ground-to-ground paths.

We make no attempt to present mathematical details or a comprehensive bibliography. Readers desiring such information are referred to a review paper by Lawrence and Strohbehn(1) containing 95 references. More recent information will be found in a review by Fante(2) and in a forthcoming book being edited by Strohbehn.(3)

The Refractive Index of the Atmosphere

The refractivity of air at optical frequencies is, to a good approximation, simply

$$N = 79 \frac{P}{T} \quad . \tag{1}$$

Here, P is the atmospheric pressure in mb and T is the temperature in K. $N \equiv (n-1)10^6$ is the deviation of the refractive index from unity in parts per million. N is roughly 290 at sea level. This approximation neglects the variation of N with wavelength and the minor effect of the variation of air density caused by the presence of water vapor. The wavelength dependence of N is about 10 percent over the optical range; the humidity dependence is less than 1 percent but its small-scale structure may sometimes be significant. These matters have been reviewed in detail by Owens(4) and by Friehe et al.(5)

Large-scale Variations and Their Optical Effects

Large-scale variations in refractivity, referring to spatial sizes greater than about a meter and durations longer than about 10 seconds, are primarily caused by the exponential decrease of pressure with height and the tendency of a convectively stable air mass to support long-lived vertical temperature gradients of large horizontal extent.

An optical ray arriving at the surface of the earth from a star suffers downward bending as it passes through the atmosphere of steadily increasing refractive index. The bending τ of the ray is known as astronomical refraction. For a grazing ray at sea level, a typical value for τ is 10 mrad (34 minutes of arc; slightly more than the apparent angular diameter of the sun). Since the optical refractive index is approximated by the dry term of the radio refractive index, the discussion of average tropospheric radio refraction given by Bean and Dutton(6), including the use of the surface value N_s to estimate τ, is directly applicable to the optical case if the humidity is taken as zero. A more elaborate procedure, including a description of the appropriate computer program and applicable to all cases, including satellite-to-satellite paths, has been published by Garfinkel.(7)

The downward bending of an optical ray also occurs on ground-to-ground paths. The curvature of the ray is caused by the vertical gradient of refractive index. Following the procedure used by Bean and Dutton,[6] in their section 2.5.3, we may differentiate the equation for refractivity to obtain, for sea-level conditions at 0° C,

$$\frac{\Delta N}{\Delta h} = .28\,\frac{\Delta P}{\Delta h} - 1.04\,\frac{\Delta T}{\Delta h} \qquad \text{(Units mentioned below)} \qquad (2)$$

If we assume that the vertical pressure gradient $\frac{\Delta P}{\Delta h}$ maintains its standard sea-level value of -12.1 mb/100 m, we can calculate the temperature gradient needed to prevent bending ($\Delta N=0$) of the ray to be -3.3 °C/100 m. Although this temperature lapse rate is much greater than for a normal atmosphere (-0.6 °C/100 m) or an adiabatic atmosphere (-1.0 °C/100 m), even greater lapse rates can occur directly above a hot surface. Such conditions give rise to a "mirage". On the other hand, if the vertical temperature gradient becomes strongly positive, as may happen above a cold surface, downward curvature of the ray can be increased to approach the curvature of the earth's surface, causing "looming". The temperature gradient needed for this is 11.8 °C/100 m. An entertaining and easily readable discussion of mirages, looming, and many other optical effects has recently been prepared by Fraser and Mach.[8]

Natural variations in the vertical temperature gradient produce hour-to-hour changes in the apparent position of a distant object. A typical difference between day and night on a 15-km path is 100 μrad, corresponding to a temperature-gradient change of 0.7 °C/100m.

Refractive-Index Turbulence

Whenever we refer to "turbulence" we shall mean refractive-index (or temperature) turbulence rather than the mechanical or velocity turbulence commonly measured with hot-wire probes. The distinction is important because, when the atmosphere is in neutral thermal stability, i.e., when the temperature lapse rate is adiabatic, strong mechanical turbulence may exist with little or no optical effect.

The most widely accepted description of the structure of mechanical turbulence appeared when Kolmogorov[9] considered the structure function D between two components of velocity, call them α,β where $\alpha,\beta = x,y,z$,

$$D_{\alpha\beta}(r) = |v_\alpha(\xi) - v_\beta(\xi+r)|^2 \qquad (3)$$

for a span r along coordinate ξ. Purely from dimensional analysis he found $D_{\alpha\beta}$ to be proportional to the two-thirds power of the separation, i.e.,

$$D_{\alpha\beta}(r) = C_{\alpha\beta}|r|^{2/3} \qquad (4)$$

where $C_{\alpha\beta}$ is a parameter depending on the components and the energy involved.

An advance of direct interest to optical propagation came when Obukhov[10] and Corrsin[11] used a similar dimensional analysis to consider the temperature fluctuations in turbulence. The structure function of this scalar parameter turns out also to obey the two-thirds law, i.e.,

$$D_T(r) = C_T^2|r|^{2/3} \qquad (5)$$

where C_T^2, the "temperature structure parameter", depends on the energy involved in the turbulence.

The structure of mechanical turbulence has been treated in detail in a number of places, e.g. Lumley and Panofsky.[12]

Pressure fluctuations in turbulent air are smoothed out with the speed of sound, and are negligible compared to temperature fluctuations. Differentiating the equation for refractivity and eliminating δP, we see that the refractive-index fluctuations are related to the temperature fluctuations, by the formula

$$\delta N = -79\,\frac{P}{T^2}\,\delta T \quad . \qquad (6)$$

Accordingly, the refractive-index structure function,

$$D_N(r) = C_N^2|r|^{2/3} = \left[\frac{79\,P}{T^2}\right]^2 C_T^2|r|^{2/3} \qquad (7)$$

where $C_N^2 = C_n^2 \times 10^{12}$ and C_n^2 is the refractive-index structure parameter. The mean value of C_N^2 tends to decrease with height above the ground, excepting a slight peak that is commonly present near the tropopause where wind shear abounds. Local variations in C_N^2 exceeding a factor of fifty seem always to be present. A model for the height dependence of C_n^2 has been formulated by Hufnagel.(13) Measurements, taken with airplanes and balloons, have been published by Tsvang,(14) by Lawrence, Ochs and Clifford,(15) and by Bufton et al.(16)

Near the ground, C_N^2 varies significantly from day to night, being primarily controlled in the daytime by the unstable convection that results from solar heating of the ground. Fig. 1 shows a typical daily variation of C_n measured with fine-wire resistance thermometers 2 m above the ground. The minima near sunrise and sunset occur when the ground temperature equals that of the air because, as mentioned earlier, if air with uniform temperature (strictly speaking, with an adiabatic lapse rate) is mixed by mechanical turbulence no optical effects are produced. At night, the atmosphere tends to be stratified in layers of different temperature and wind-driven turbulent mixing produces measurable levels of C_N.

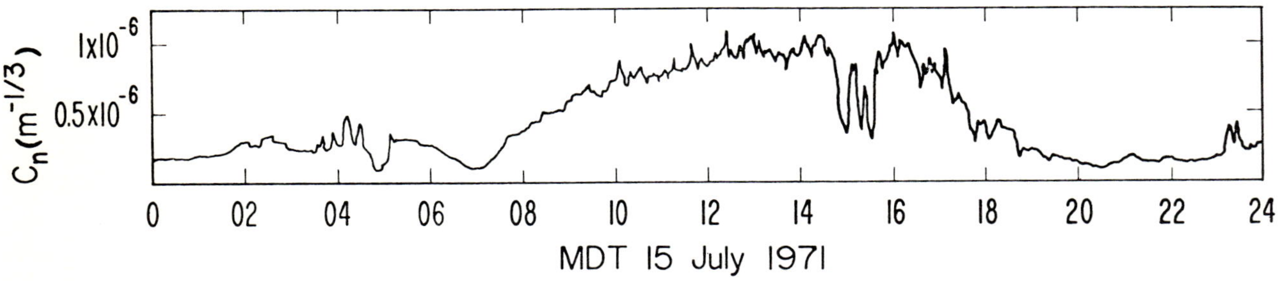

Fig. 1. Typical diurnal variation of the strength of refractive-index turbulence near the ground. The daytime peak results from heating of the ground; the dips at 15 hours are caused by clouds passing in front of the sun.

Optical Effects of Refractive-Index Turbulence

Phase and Amplitude Distortion of the Wave Front

Referring to Fig. 2, let us consider the behavior of a plane wave front A such as might arrive from a star, as it travels through the turbulent atmosphere. Immediately after passing through a region of irregular refractive index, the wave front B has been distorted. Since absorption and wide-angle scattering are negligible in the clear atmosphere, the energy density of the wave front B is still uniform and equal to its free-space value. Thus an ordinary square-law detector placed at B would be unaffected by the wave-front distortions and incapable of measuring them. The distortions can, of course, be measured by a phase-sensitive detector such as an interferometer or a telescope.

As the wave progresses from B toward C, the various portions of the distorted wave front travel in slightly different directions and eventually begin to interfere. The interference is equivalent to a redistribution of energy in the wave and causes intensity fluctuations (scintillations) that can be detected by a square-law detector. On the way from B to C, the wave passes through additional refractive-index irregularities and so suffers additional phase perturbations. These new irregularities are, however, relatively ineffective in producing scintillations at C.

Let us examine the criteria that determine which of the turbulence irregularities along a line of sight are most effective in producing intensity fluctuations. In Fig. 3, consider an irregularity of diameter ℓ at an arbitrary point Z on the line of sight between the plane wave and the receiver at L. The irregularity can be fully effective in producing intensity variations only if the extreme ray paths AL and ZL differ in length by at least half a wavelength, i.e., the irregularity must be at least equal in size to the first zone of a Fresnel zone plate situated at Z. This minimum effective size is, in fact, the optimum size for the irregularity. Larger irregularities at the same point are ineffective because they do not diffract light through a large enough angle to reach the observer. While it is true that smaller irregularities produce intensity fluctuations at points closer than L and that these fluctuations persist in modified form until the wave reaches L, such smaller irregularities are relatively ineffective because of the steep increase in the Kolmogorov spectrum of atmospheric turbulence with irregularity size.

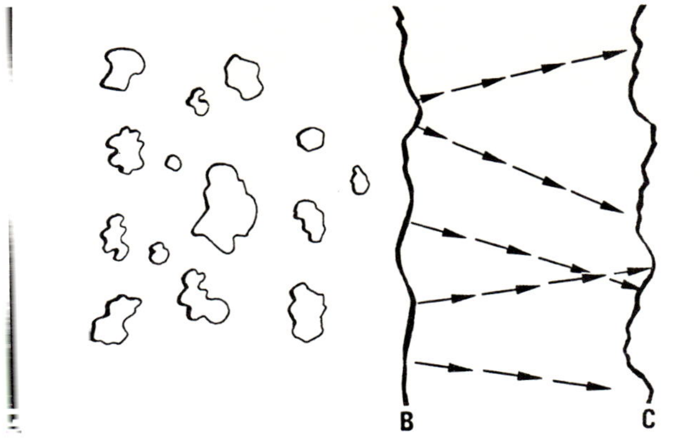

Fig. 2. The effect of atmospheric irregularities on a plane wave A. At B there are only phase distortions of the wave front; at C there are both amplitude and phase distortions.

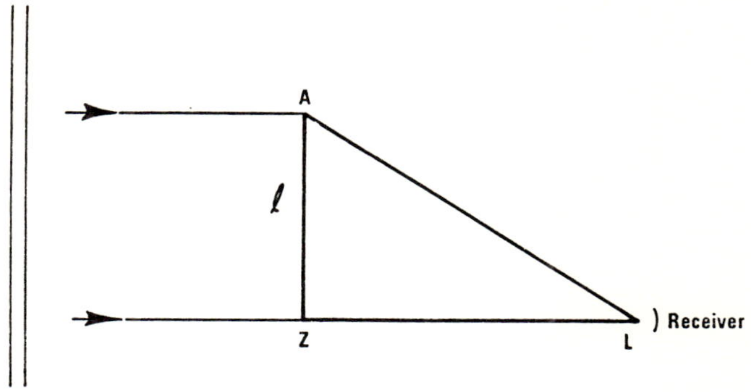

Fig. 3. The size ℓ of the irregularities most effective in producing scintillations is such that distance AL exceeds distance ZL by one-half wavelength.

From the geometry of Fig. 3, we see that the diameter of the most effective irregularity at height h is $\ell = \sqrt{\lambda h}$, where λ is the wavelength of the light. As a numerical example, for visible light of wavelength 0.5 μm and for a height of 10 km, ℓ = 7 cm.

If we assume, for the moment, that the turbulence is uniformly distributed along the path between the top of the atmosphere and the receiver, and that it has a Kolmogorov spectrum so that the large irregularities are more intense than the small, it is clear that the mean-square fluctuation of refractive index attributable to irregularities of optimum size varies systematically along the path. There is, therefore, a weighting function that expresses the relative effectiveness of turbulence in producing intensity fluctuations as a function of path position. It is clear that this weighting function must be a maximum at the top of the atmosphere (or at the middle of the path for a ground-to-ground path with a point source) and must decrease monotonically to zero at the receiver.

It is also clear that, for the plane-wave case of starlight, the intensity pattern at the receiver has the same scale size as the irregularities that produced it. Thus, in our numerical example, turbulence at a height of 10 km produces an intensity pattern on the ground that has a predominant size of 7 cm while turbulence at lower heights produces smaller scales.

The arguments given above assume simple superposition of the effects of irregularities at various points along the path. When the integrated turbulence on the entire path becomes too large, this assumption fails and the scintillations "saturate", i.e., the strength of the scintillations is no longer proportional to the strength of the turbulence that produced them. When saturation exists, the most effective portion of the path moves toward the receiver (or, for ground-to-ground paths with a point source, toward the ends of the path) and the scale sizes in the pattern are affected accordingly.

In summary, distortions in the phase of a plane wave from a star traveling through a clear uniformly turbulent atmosphere are produced equally by turbulence at all locations. Because of the Kolmogorov spectrum, the large-scale eddies in the turbulence produce greater phase shifts than do the smaller scales. The amplitude, or intensity fluctuations in the wave front are produced predominantly by the turbulence high in the atmosphere, and the sizes of the pattern vary as the square root of the distance to the turbulence.

It is convenient to describe the magnitude of the phase fluctuations in terms of the phase structure function, $D_\phi(r)$, which is simply the mean square phase differences between two points a distance r apart on a plane parallel to the mean wave front. The Kolmogorov spectrum of turbulence produces wave front distortions that have a particular form of phase structure function, viz.,

$$D_\phi(r) \propto r^{5/3} . \qquad (8)$$

Thus, for waves that have passed through natural turbulence in the atmosphere, the magnitude and nature of the phase fluctuations can be completely specified by a single number, r_0, known as the phase coherence length, which is the value of r for which $D_\phi(r)=1$.

For starlight at the ground

$$D_\phi(r) = \frac{28.8}{\lambda^2} r^{5/3} \int_0^\infty C_n^2 \, dh \tag{9}$$

in SI units, and r_o is of the order of 5 to 15 cm.

The strength of scintillations, on the other hand, is described in terms of the contrast between the light and dark areas. The most common measure of this contrast is the "log-amplitude variance"

$$\sigma_\chi^2 = \frac{1}{4}\left\langle\left[\log_e \frac{I}{I_o}\right]^2\right\rangle \tag{10}$$

where I is the instantaneous intensity (or irradiance) of the wave and I_o is its long-term average. The pointed brackets denote averaging. Another frequently used measure of the strength of scintillations is the "scintillation index"

$$S_I = \frac{[<(I-I_o)^2>]^{\frac{1}{2}}}{I_o} \; . \tag{11}$$

For scintillations produced by the Kolmogorov turbulence in the atmosphere, the logarithm of the intensity has a normal or Gaussian probability distribution and the relation between log-amplitude variance and scintillation index is

$$\sigma_\chi^2 = \frac{1}{4} \log_e(1+S_I{}^2) \tag{12}$$

Notice that, unlike phase fluctuations, the scintillations are measured by quantities that give no indication of the scale size of the pattern. This is because the pattern size is usually determined by the distance to the turbulent layers and is independent of the strength of the turbulence. An exception occurs when the scintillations saturate, i.e., when $\sigma_\chi^2 \geq 0.6$, and the Fresnel-zone sized irregularities begin to disappear relative to larger and smaller scales. For starlight, this happens only for stars near the horizon.

The log-amplitude variance is related to the strength of turbulence C_n^2 throughout the atmosphere by the formula

$$\sigma_\chi^2 = \frac{4.78}{\lambda^{7/6}} \int_0^\infty C_n^2 \, h^{5/6} \, dh \tag{13}$$

in SI units. The factor $h^{5/6}$ in this formula shows that the turbulence at great heights is more effective in producing scintillations than is the turbulence near the ground.

Effect of Wave-Front Distortions on a Stellar Image

The phase and intensity distortions of the wave front each affect the ability of a telescope to form a point image of a star, though in different ways. Considering first the phase distortions, we must discuss separately the effect of scale sizes larger than the telescope aperture diameter d and those smaller than d. If we call the mean-square variations of phase difference over the aperture due to the large scale sizes $\Delta\phi_\ell^2$ and the corresponding variations due to small sizes $\Delta\phi_s^2$, we can write

$$\Delta\phi_s^2 + \Delta\phi_\ell^2 = D_\phi(d) \tag{14}$$

The phase variations $\Delta\phi_\ell$ larger than d simply change the position of the stellar image in the focal plane of the telescope. Thus, for exposure times short compared with the time it takes for wind to blow across the diameter of the telescope, the quality of the image is unaffected by these large phase irregularities. On longer exposures, the image wanders through an angle λ/r_o, causing a long-exposure smearing of the image. This long-exposure smearing angle is independent of telescope diameter as long as $d>r_o$.

The phase variations $\Delta\phi_s$ smaller than the telescope aperture act as independent prisms of diameter r, each forming a separate image that is spread by a diffraction angle λ/r. These become important only as $D_\phi(r)$ approaches unity, and the integrated effect of all these small phase irregularities produces the "short-exposure" spreading which is of the order of λ/r_o^{ST}. The short-exposure phase coherence length r_o^{ST} is defined in the same manner as r_o, except that only phase irregularities smaller than the aperture diameter are involved. Yura[17] has shown that

$$r_o^{ST} \approx r_o[1+0.37(r_o/D)^{1/3}] \tag{15}$$

which is 10 to 40 percent greater than r_o for telescopes of modest diameter. We see that, as far as the formation of a small stellar image is concerned, there is no advantage to having a telescope diameter larger than r_o^{ST}.

The scintillations may also degrade the image if the scintillation index, as measured by a small-aperture receiver, is greater than about 0.2. Since the scintillations have a predominant size equal to a Fresnel zone $\sqrt{\lambda h}$ (where h is the height of the highest scintillation-producing layer in the atmosphere), the angle through which these random apodizers can spread the image is of the order of

$$\frac{\lambda}{\sqrt{\lambda}} = \sqrt{\lambda/h} \quad . \tag{16}$$

We see that scintillations cause significant image degradation relative to that caused by phase fluctuations only when $r_o > \sqrt{\lambda h}$. This, of course, depends on the relative strength of the turbulence in the upper atmosphere compared with that near the ground just above the observatory. The intensity of the blurred image caused by scintillation is always less than that caused by phase fluctuations, the ratio never exceeding about 0.2.

References

1. Lawrence, R. S. and Strohbehn, J. W., "A Survey of Clear-Air Propagation Effects Relevant to Optical Communications,", Proc. IEEE (Special Issue) 58:10, 1523-1545, 1970.

2. Fante, R. L., "Electromagnetic Beam Propagation in Turbulent Media," Proc. IEEE 63:12, 1669-1692, 1975.

3. Strohbehn, J. W. (ed), "Topics in Applied Physics; Laser Beam Propagation Through the Atmosphere," Springer-Verlag (to be published).

4. Owens, J. C., "Optical Refractive Index of Air: Dependence on Pressure, Temperature, and Composition," Appl. Opt. 6:1, 51-59, 1967.

5. Friehe, C. A., LaRue, J. C., Champagne, F. H., Gibson, C. H., and Dreyer, G. F., "Effects of Temperature and Humidity Fluctuations on the Optical Refractive Index in the Marine Boundary Layer,", J. Opt. Soc. Am. 65:12, 1502-1511, 1975.

6. Bean, B. R. and Dutton, E. J., Radio Meteorology, NBS Monograph 92, U. S. Government Printing Office, 1966. (also, Dover Publications, 1968.)

7. Garfinkel, B., "Astronomical Refraction in a Polytropic Atmosphere," Astron. J. 72:2, 235-254, 1967.

8. Fraser, A. B. and Mach, W. H.,"Mirages," Sci. Amer. 234:1, 102-111, 1976.

9. Kolmogorov, A., Turbulence, Classic Papers on Statistical Theory, S. E. Friedlander and L. Topper, Eds., New York, 1941: Interscience p. 151, 1961.

10. Obukhov, A. M., "Structure of the Temperature Field in a Turbulent Flow", Izv. Akad. Nauk, SSSR. Ser. Geograf. Geofiz. 13:58, 1949.

11. Corrsin, S., "On the Spectrum of Isotropic Temperature Fluctuations in an Isotropic Turbulence", J. Appl. Phys. 22:, 469-473, 1951.

12. Lumley, J. L. and Panofsky, H. A., The Structure of Atmospheric Turbulence, John Wiley & Sons, 1964.

13. Hufnagel, R. E., "Variations of Atmospheric Turbulence," Optical Society of America Topical Meeting on Optical Propagation Through Turbulence, Boulder, Colorado July 1974.

14. Tsvang, L. R., "Some Characteristics of the Spectra of Temperature Pulsations in the Boundary Layer of the Atmosphere," Izv. Geophys. Ser., No. 10, 1594-1600, 1963.

15. Lawrence, R. S., Ochs, G. R. and Clifford, S. F., "Measurement of Atmospheric Turbulence Relevant to Optical Propagation," J. Opt. Soc. Am. 60:6, 826-830, 1970.

16. Bufton, J. L, Minott, P. O., Fitzmaurice, M. W., and Titterton, P. J., "Measurements of Turbulence Profiles in the Troposphere," J. Opt. Soc. Am. 62:9, 1068-1070, 1972.

17. Yura, H. T., "Short-Term Average Optical-Beam Spread in a Turbulent Medium," J. Opt. Soc. Am. 63:5, 567-572, 1973.

AN ELEMENTARY DERIVATION OF PHASE FLUCTUATIONS OF AN OPTICAL WAVE IN THE ATMOSPHERE

H. T. Yura
Electronics Research Laboratory
The Aerospace Corporation
P.O. Box 92957
Los Angeles, California 90009

Abstract

This paper presents an elementary derivation of the basic phase statistics of an optical wave (both plane and spherical) propagating in a turbulent medium. The results are general in that they apply to an arbitrary passive spatially homogeneous turbulent medium. Explicit application of the analysis is directed to the atmosphere where we obtain the well known results of Tatarskii and others. The present approach provides an intuitive appreciation of the physical behavior of optical wave propagation through a medium which exhibits a spatially random index of refraction. As such, it is complementary to the rigorous mathematical derivations that appear in the literature.

Introduction

The study of the statistics of optical wave propagation in turbulent media has received increased interest in recent years. This interest is due primarily to the advent of laser systems operating in the atmosphere and ocean. For instance, the design and development of optical imaging and radar systems operating in the atmosphere should be based on the results of the analysis of an optical wave field propagating in turbulent media. Other areas of interest where the interaction of an optical beam with random media is important include astronomy, laser communication and remote sensing. Therefore, it is important to have a theory for predicting the nature of the propagation of a beam of light in a random medium. During the last fifteen years a great deal of progress was made on this problem(1-5). In particular, useful expressions were obtained for essentially all optical wave quantities of interest. In particular, results have been obtained for beam irradiance, scintillation, coherence, etc., as a function of propagation distance L, optical wave number k (equal to 2π divided by the optical wavelength λ), and the parameters that describe the turbulent medium. These results were shown to be in fairly good agreement with experiment. The agreement between experiment agrees very well with theory in the regime of weak fluctuations. In the regime of strong fluctuations, that is, in the regime where the variance of irradiance saturates, experiment and the results of recent theoretical efforts(6, 7, 8) appear to agree well qualitatively, and quantitatively(9).

Although a great body of literature exists on this subject (see for example Refs. 4 and 5 for excellent reviews that list several hundred references), much of the analysis is mathematically complex, inhibiting an understanding of the underlying physical optic effects. In this paper we will derive the basic phase statistics which play a central role in propagation theory using a minimum of mathematics. Rather, we will employ physically intuitive arguments that will enable us to obtain all of the plane and spherical wave statistics (to within a numerical multiplicative factor of the order unity). We believe that a physical approach is useful for a complete understanding of wave propagation in turbulent media and should be used in a complementary fashion along with the rigorous derivations given in the literature.

First we will outline the model assumed for the index of refraction fluctuations in the atmosphere and then give a physical derivation of phase statistics. Random fluctuations of temperature in the earth's atmosphere result in fluctuations in the index of refraction. These fluctuations are in general functions of the position $\underline{r}$ and time t so that the index of refraction n can be written as

$$n(\underline{r}, t) = 1 + n_1(\underline{r}, t) , \tag{1}$$

where n_1 is the fluctuation in the index of refraction. For clear-air atmospheric turbulence, we have that $|n_1|^1 << 1$, and $\langle n_1 \rangle = 0$, where angular brackets denote the ensemble average. It is also possible to assume that the temporal dependence of n_1 is mainly due to a net transport of the inhomogenieties of the medium as a whole past the line of site (e.g., due to atmospheric winds) so that $n_1(\underline{r}, t) = n_1(\underline{r} - v(\underline{r})t)$, where $v(\underline{r})$ is the local "wind" velocity. This assumption is known as Taylor's frozen-flow hypothesis and appears to hold in most practical cases of interest. Here we are primarily interested in obtaining spatial statistics and so we suppress the explicit temporal behavior of n_1.

For simplicity, we assume that the turbulence condition are both homogeneous and isotropic. For atmospheric turbulence we employ the much used Kolmogorov model according to which within a particular range of separations between $\underline{r}_1$ and $\underline{r}_2$ that

$$\langle [n(\underline{r}_1) - n(\underline{r}_2)]^2 \rangle = C_n^2 |\underline{r}_1 - \underline{r}_2|^{2/3}, \qquad \ell_o << |\underline{r}_1 - \underline{r}_2| << L_o \tag{2}$$

where ℓ_o and L_o are called the inner and outer scales of turbulence, respectively, and C_n is called the index structure constant. In the atmosphere, ℓ_o is of the order a few millimeters to a centimeter and, for horizontal propagation in the lower atmosphere, L_o is of the order the height above ground.

For turbulent eddy sizes, of characteristic length r, within the inertial subrange (i.e., $\ell_o \ll r \ll L_o$), the Kolmogorov "2/3" law implies that to a numerical multiplicative factor of the order unity the mean square index fluctuation associated with the scale size r increases as the two-thirds power of that scale size. That is,

$$\langle n_1^2(r) \rangle \sim C_n^2 \, r^{2/3} \,, \qquad \ell_o \ll r \ll L_o \,. \tag{3}$$

For separations larger than L_o, the mean square index fluctuation does not increase with increasing separation, rather it levels off to a value of the order $C_n^2 L_o^{2/3}$. Conversely, for separations small compared with ℓ_o, friction effects due to viscosity result in a very rapid decrease in the index fluctuations and for many applications can be taken as being equal to zero (ℓ_o is frequently smaller than any length of interest in a propagation problem resulting in negligible contributions due to the smallest scale sizes).

In summary, the present analysis is restricted to a weakly inhomogeneous medium with characteristic scale lengths much greater than the optical wavelength. Furthermore, it is assumed that the characteristics of the medium do not change appreciably during an oscillation of the optical field, for otherwise frequency spreading (Doppler effects) becomes important. At ir and visible wavelengths this condition is satisfied in the atmosphere. The electromagnetic field considered has a time dependence given by $\exp(i\omega t)$. The time dependent wave equation is replaced by the Helmholtz equation for a random inhomogeneous medium. The electrical conductivity and magnetic permeability of the medium are taken to be 0 and 1, respectively. Finally, although only the case of a scalar (plane and spherical) wave propagating in a spatially homogeneous and isotropic turbulent medium is discussed, the extension to the case of inhomogeneous turbulence conditions is straightforward.

Phase Statistics

Consider a plane wave impinging on a random medium that occupies the half-space $z \geq 0$. Normalizing the amplitude of the wave to unity in the absence of turbulence and suppressing the time dependence we have that the optical wave at propagation distance L can be expressed as e^{ikL}. In the presence of the random medium we express the optical wave function as $U(\underline{r}) = A \exp(ikL + i\phi)$, where, in general, both the amplitude perturbation A and phase perturbation ϕ are random functions of position $\underline{r}$ in the medium. Here we assume that A is constant and consider phase fluctuations only.

The mutual coherence function (MCF), which is defined as the cross-correlation function, i.e., second statistical moment, of the complex field in a plane perpendicular to the mean direction of propagation, is important for a number of reasons (e.g.; it describes the loss of coherence of an initially coherent wave propagating in the medium). Neglecting amplitude effects the MCF is given by

$$M \equiv \langle U(\underline{r}_1)\, U^*(\underline{r}_2) \rangle = \langle \exp\{ i[\phi(\underline{r}_1) - \phi(\underline{r}_2)] \} \rangle \tag{4}$$

where $\underline{r}_{1,2} = (\underline{\rho}_{1,2}, L)$, and $\underline{\rho}_{1,2}$ is a two-dimensional vector in a plane transverse to the optic axis at propagation distance L (see Figure 1).

In order to calculate the ensemble average indicated in Eq. (4), it is convenient to first obtain an expression for $\phi(\underline{r})$. To this end, we note that since $\lambda \ll$ turbulent scale sizes of the medium, geometrical optics should yield a reasonable estimate for $\phi(r)$ [1,2]. We will show that geometrical optics is entirely adequate to obtain the second order phase statistics, first obtained by Tatarskii [1,2] and Chernov [3].

According to geometrical optics, the phase perturbation ϕ resulting from the fluctuation in the index of refraction n_1

$$\phi = k \int_{\text{path}} n_1(s)\, ds \tag{5}$$

where s is position along the ray path. For plane waves the paths are approximately straight lines parallel to the z-axis, as indicated in Fig. 1. As long as the propagation distance L is much larger than the integral correlation scale of n_1 (as it will be for essentially all cases of practical interest), Eq. (5) indicates that we add together a large number of uncorrelated terms to obtain ϕ. Hence because of the central limit theorem, we conclude that ϕ will be normally distributed, independent of the distribution of n_1. A normal distribution is completely characterized by two parameters, its mean and variance. Since $\langle n_1 \rangle = 0$, it follows directly from Eq. (5) that $\langle \phi \rangle = 0$. The cross-correlation function of the phase is obtained from Eq. (5) as

$$\begin{aligned} B_\phi(|\underline{\rho}_1 - \underline{\rho}_2|) &= \langle \phi(\underline{\rho}_1)\, \phi(\underline{\rho}_2) \rangle = k^2 \left\langle \left(\int n_1(s_1)\, ds_1\right) \left(\int n_1(s_2)\, ds_2\right) \right\rangle \\ &= k^2 \iint ds_1\, ds_2 \langle n_1(s_1)\, n_1(s_2) \rangle = k^2 \iint ds_1\, ds_2\, B_n(|s_1 - s_2|) \end{aligned} \tag{6}$$

where we have introduced the index autocorrelation function B_n. Since the medium is assumed statistically stationary and isotropic, it follows that B_n is a function of the magnitude of the difference of the two points s_1 and s_2.

From Fig. 1 we find that

$$|s_1 - s_2| = \left(z_{12}^2 + \rho^2\right)^{1/2} \tag{7}$$

where

$$\rho = |\underline{p}_1 - \underline{p}_2|$$

and

$$z_{12} = z_1 - z_2 \tag{8}$$

Note that the mean square phase, $\phi_o^2 \equiv \langle \phi^2 \rangle$, is obtained from Eq. (6) as $B_\phi(0)$. Thus, in general, the phase is normally distributed with a zero mean value and with a second moment given formally by Eq. (6).

Returning now to the evaluation of the ensemble average indicated in Eq. (4). It can be shown that if the random function f is normally distributed,

$$\langle e^f \rangle = \exp\left[\langle f \rangle + \frac{1}{2}\langle [f - \langle f \rangle]^2 \rangle\right] \tag{9}$$

Applying Eq. (9) to Eq. (4) with $f = i[\phi(r_1) - \phi(r_2)]$ yields

$$M = \exp[-\frac{1}{2} D_\phi(\rho)] \tag{10}$$

where

$$D_\phi(\rho) \equiv \langle [\phi(\underline{r}_1) - \phi(\underline{r}_2)]^2 \rangle \tag{11}$$

is called the phase structure function in the literature [1,2], and $\underline{\rho} = \underline{r}_1 - \underline{r}_2$. Expanding the square in Eq. (11) and invoking stationarity yields that

$$M = \exp\left[-\frac{1}{2} D_\phi(\rho)\right] = \exp\left\{-\phi_o^2[1 - b_\phi(\rho)]\right\} \tag{12}$$

where ϕ_o^2 is the mean square phase fluctuation at a point and $b_\phi(\rho)$ is the normalized phase correlation function defined as

$$b_\phi(\rho) \equiv \frac{\langle \phi(\underline{r}_1)\,\phi(\underline{r}_2) \rangle}{\langle \phi^2 \rangle} = \frac{B_\phi(\rho)}{\phi_o^2} \tag{13}$$

From its definition it is seen that $b_\phi(0) = 1$. Furthermore, for sufficiently large separations ρ (i.e., large compared to the outer scale of turbulence) the ray that arrives at $\underline{p}_1$ is statistically independent of the ray arriving at $\underline{p}_2$ from which it follows that $b_\phi \to 0$. That is, for large ρ, $\langle U(\underline{r}_1)\, U^*(\underline{r}_2) \rangle \to \langle U(r_1) \rangle \times \langle U^*(r_2) \rangle = \langle U \rangle^2 = \exp(-\phi_o^2)$. The MFC decreases monotonically from $M(0) = 1$ to $M(\infty) = \exp(-\phi_o^2)$.

While it is a straighforward numerical procedure to compute M directly for a given functional form of the index correlation function, it is useful, indeed preferable, to have approximate formulas for the mutual coherence function for estimating the coherence at various ranges.

We assume that the medium is composed of a continuum of independent turbulent eddies from a minimum scale size ℓ_o to a maximum scale size L_o. Consider for a moment an eddy of scale size a. We now calculate the contribution to M from this eddy size. Since the eddies are independent the resulting MCF due to all eddies will be the product of the MCF due to a single scale size over all scale sizes. A physically adequate and mathematically convenience form for B_n is a gaussian. As we shall show, the gaussian form is sufficient for a complete determination of second order phase statistics. Indeed, any physically correct form for B_n will suffice; however the gaussian is particularly convenient mathematically. Thus, consider that

$$B_n(\rho) = \langle n_1^2 \rangle \exp(-\rho^2/a^2), \tag{14}$$

where $\langle n_1^2 \rangle$ is the mean square fluctuation in the index of refraction and a is the eddy scale size of interest. Substituting Eq. (14) into Eq. (6), making use of Eq. (7) and (8) yields, for plane waves,

$$B_\phi(\rho) = k^2 \langle n_1^2 \rangle e^{-\rho^2/a^2} \int_0^L dz_1 \int_0^L dz_2\, e^{-(z_1 - z_2)^2/a^2} \tag{15}$$

Changing variable from z_1, z_2 to $\eta = z_1 - z_2$ and $\xi = (z_1 + z_2)/2$, and noting that L >> a allows one to express B_ϕ as

$$B_\phi(\rho) \simeq k^2 \langle n_1^2 \rangle L e^{-\rho^2/a^2} \int_0^\infty d\eta\, e^{-\eta^2/a^2} = \phi_o^2 e^{-\rho^2/a^2} \tag{16}$$

where

$$\phi_o^2 = \frac{\sqrt{\pi}}{2} k^2 \langle n_1^2 \rangle aL. \tag{17}$$

Indeed, the phase variance ϕ_o^2 for a single scale size a can be shown on physical grounds to always be of order $k^2 \langle n_1^2 \rangle aL$. Consider an eddy of size a. The phase change of a ray passing through this eddy is of the order $k n_1 a$. The mean square phase change per eddy is of the order $k^2 \langle n_1^2 \rangle a^2$. For a total path length L, there are of the order L/a eddies traversed by the ray. Thus, for independent eddies, the mean square phase change is equal to the mean square phase change per eddy multiplied by the number of eddies traversed by the ray. That is,

$$\phi_o^2 \sim (k^2 \langle n_1^2 \rangle a^2)(L/a) = k^2 \langle n_1^2 \rangle aL.$$

From Eqs. (13) and (16) we deduce that the normalized correlation coefficient for plane waves is

$$b_\phi(\rho) = \frac{B_\phi(\rho)}{\phi_o^2} = e^{-\rho^2/a^2} \tag{18}$$

Thus from Eqs. (10), (12), and (16) we find that

$$M_a(\rho) = \exp\left[-\phi_o^2(a)\left(1 - e^{-\rho^2/a^2}\right)\right] \tag{19}$$

where, in general,

$$\phi_o^2(a) \sim k^2 \langle n_1^2(a) \rangle aL\ , \tag{20}$$

and is given explicitly for a gaussian index autocorrelation function by Eq. (17). Note that ϕ_o^2 can be written as $2z/z_c$ where the extinction length $z_c \sim [k^2 \langle n_1^2 \rangle a]^{-1}$ is the distance where the coherent part of the optical field is reduced by 1/e from its value at z = 0.

Now for a continuum of eddy sizes distributed according to the Kolmogorov "2/3" law, we obtain from Eqs. (3) and (20) that

$$\phi_o^2(a) \sim k^2 C_n^2 L a^{5/3} \tag{21}$$

Examination of Eq. (21) reveals that the largest eddies are most effective in contributing to the total phase variance at a point. In fact, for a distribution of independent eddies we have

$$\phi_o^2 \sim \sum_{\ell_o}^{L_o} (k^2 C_n^2 L) a^{5/3} \sim k^2 C_n^2 L L_o^{5/3}\ ,$$

in agreement in order of magnitude with the results of the rigorous calculation of Tatarskii and others (the numerical multiplicative constant is approximately 0.39). Furthermore, the resulting MCF of a distribution of independent eddies is

$$M(\rho) = \prod_{\ell_o}^{L_o} M_a(\rho) = \exp\left\{-\sum_{\ell_o}^{L_o} \phi^2(a)\left[1 - e^{-\rho^2/a^2}\right]\right\} \tag{22}$$

where, for the Kolmogorov "2/3" law, $\phi_o^2(a)$ is given by Eq. (21). Examination of Eq. (22) reveals that there are three distinct range values for ρ.

Range 1: $\rho \gg L_o$. Over this range $b_\phi(\rho) = 0$ since ρ is larger than the outer scale of turbulence. It follows from Eq. (22) that for $\rho \gg L_o$

$$M \to \exp\left\{-\sum_{\ell_o}^{L_o} \phi_o^2(a)\right\} \sim \exp\left[-\phi_o^2(L_o)\right] = \exp\left[-k^2 C_n^2 L L_o^{5/3}\right]. \tag{23}$$

Thus, in this regime the MCF is a constant independent of ρ [and equal to $\langle U \rangle^2 = \exp(-\langle \phi^2 \rangle)$].

Range 2: $\ell_o \ll \rho \ll L_o$. This range of ρ values is called the inertial subrange and is the most interesting case in practice. In the inertial subrange an examination of Eqs. (21) and (22) reveals that the argument of the outer exponential in Eq. (22) increases proportional to $a^{5/3}$ for $a < \rho$, reaches a maximum for $a \sim \rho$, and decreases proportional to $a^{-1/6}$ for $a > \rho$. Thus the most effective eddy scales in the inertial subrange are those of scale lengths of the order the separation of interest. As a result we obtain that for $\ell_o << \rho << L_o$

$$M(\rho) = \exp[-(\rho/\rho_o)^{5/3}] \quad , \tag{24}$$

where, the "coherence length" ρ_o is given by

$$\rho_o \sim (k^2 C_n^2 L)^{-3/5} \quad , \tag{25}$$

This result agrees, to within a numerical multiplicative factor of the order unity, with the results of Tatarskii [1,2]. In the inertial subrange the phase structure function is proportional to $\rho^{5/3}$ and for almost all cases of practical concern is the regime of separations of interest.

Range 3: $\rho << \ell_o$. In this regime ρ is much smaller than all eddy scale sizes from which it follows that the inner exponent in Eq. (22) may be expanded for each a as $\exp(-\rho^2/a^2) \cong 1 - \rho^2/a^2$. (In general it can be shown [2] that the underlying structure function has the property for $\rho \rightarrow 0$ that $D(\rho) \sim O(\rho^2)$. For the case of homogeneous and isotropic turbulence this implies that $B \sim 1 - O\rho^2$.) As a result, we obtain from Eqs. (21) and (23) that the argument of the outer exponent in Eq. (22) is proportional to $a^{-1/6}$. Thus for $\rho << \ell_o$, the main contribution to the sum indicated in Eq. (22) are for eddies of the order ℓ_o, the inner scale of turbulence. Thus, for $\rho << \ell_o$, we obtain the estimate that

$$M(\rho) = \exp[-(k^2 C_n^2 L/\ell_o{}^{1/3})\,\rho^2] \tag{26}$$

In this regime of separations the phase structure function is a quadratic function of ρ. The results obtained in Eq. (26) are in agreement, to within a numerical multiplicative factor of the order unity, with the results of Tatarskii.

As alluded to previously, the inertial subrange is the regime of interest in practice. We note from Eq. (25) that the restriction $\ell_o << \rho << L_o$ is equivalent to the propagation distance condition that $z_c << L << z_i$, where the $z_c \sim (k^2 C_n^2 L_o^{5/3})^{-1}$ is the distance at which the coherent part of the field is reduced by 1/e from its value at L = 0, and $z_i \sim (k^2 C_n^2 \ell_o{}^{5/3})^{-1}$ is the propagation distance at which the coherence length, ρ_o, is equal to the inner scale of turbulence. For example, for ℓ_o = 1 mm, L_o = 1 m, λ = 0.5 μ, and $C_n^2 = 10^{-14}\, m^{-2/3}$ (fairly strong turbulence conditions), we obtain that $z_c \sim$ 1 m, and $z_i \sim$ 100 km, while for $C_n^2 = 10^{-16}$ (fairly weak turbulence conditions) and the same value for the other parameters we obtain that $z_c \sim$ 100 m, and $z_i \sim$ 10,000 km. In many applications of interest the distance z_i is frequently much larger than the propagation distance of interest. For L less than z_i, the inner scale does not enter into the analysis (i.e., $\rho_o > \ell_o$ and we can approximate the plane wave MCF by the convenient form ($L < z_i$)

$$M(\underline{\rho}) \simeq \exp\left\{-\phi_o^2\left[1 - \exp\left[-\frac{1}{\phi_o^2}\left(\frac{\rho}{\rho_o}\right)^{5/3}\right]\right]\right\} = \exp\left\{-\phi_o^2\left[1 - \exp\left[-(\rho/L_o)^{5/3}\right]\right]\right\}$$

$$\rightarrow \exp\left[-(\rho/\rho_o)^{5/3}\right], \quad \rho << L_o \tag{27}$$

$$\rightarrow \exp\left[-\phi_o^2\right], \quad \rho \gtrsim L_o \quad .$$

Next we indicate for spherical waves the modification to the plane wave results. Consider a monochromatic point source located at the origin and we week to determine the resulting MCF for separation ρ at propagation distance L. With reference to Fig. 2, the derivation that led to Eqs. (6) - (8), and noting that the correlation scales of the medium are much less than the propagation distance L we find for spherical waves that

$$B_\phi \cong k^2 \int_0^L dz_1 \int_0^L dz_2\, B_n(|s_1 - s_2|) \tag{28}$$

where

$$|s_1 - s_2| = [z_{12}^2 + (t\rho)^2]^{1/2}$$

and $$t = \frac{z_1 + z_2}{2L} \tag{29}$$

The modification for spherical wave propagation is geometrical in nature. The dependence on the quantity t in Eq. (29) reflects the spherical expansion of the wave as it propagates through the medium. We note from Eqs. (28) - (29) that the phase variance for spherical waves is identical to that obtained for plane waves, as it should. The resulting MCF for spherical waves is

$$M_s(\rho) = \exp\left\{-\phi_0^2\left[1 - \int_0^1 b_\phi(\rho t)\,dt\right]\right\} \quad (30)$$

where b_ϕ is the plane wave normalized phase correlation coefficient. Thus, the spherical wave coherence function is obtained directly from the plane wave results, as indicated above. For uniform turbulence conditions, it is seen that the estimates of M obtained for the three range values for ρ discussed above apply to spherical waves. The only difference being, for ranges 2 and 3, a numerical multiplicative factor (the additional factor multiplying $\rho^{5/3}$ and ρ^2 being 3/8 and 1/3, respectively).

The spherical wave coherence length is in general, greater than the corresponding plane wave coherence length, as expected intuitively from geometrical expansion. Note that, in contrast to plane waves, for inhomogeneous turbulence conditions the spherical wave coherence length is dependent on source location (e.g., spherical wave distortions for propagating up and down through the atmosphere are quite different implying corresponding differences in the ultimate achievable resolution).

The spherical wave mutual coherence function plays an important role in propagation theory. For example, it describes the reduction in lateral coherence between different elements of a transmitting or receiving aperture, effectively transforming them into partially coherent apertures, with the degree of coherence decreasing with increasing propagation distance through the medium. In general, the resulting mean irradiance distribution (in space for a transmitting system, or in the image plane for receiving system) is characteristics of a coherent aperture of dimension ρ_0. If ρ_0 is smaller than the diameter of the physical aperture D, the resulting radiation pattern is characteristic of that of an aperture of diameter ρ_0 rather than D. Hence, in the far field, the scattering half angle is given by the reciprocal of the product of the wave-number and the radius of the "effective" coherent aperture (which is approximately given by the smaller of $\rho_0/2$ or D/2).

By physically intuitive arguments we have obtained the basic phase statistics results obtained by Tatarskii using perturbation theory and the method of spectral expansions [1,2]. Note from Eq. (6) that the phase statistics are directly proportional to the index of refraction correlation function. The incremental contribution to B_ϕ from random inhomogeneities in a slab of thickness Δz located at a distance L is proportional to the product of the index fluctuations located in the slab, which is (assumed) independent of the properties of the optical wave. This suggests that the results derived above [which agree with the second order (in n_1) perturbation theory results] should be valid for arbitrary propagation distance L. In particular, they should be valid in the regime where the normalized variance of irradiance saturates. In the so-called saturation regime, the method of smooth perturbation applied to amplitude fluctuations is not valid. The incremental contribution to amplitude fluctuations from eddies in a slab of thickness Δz, in contrast to phase fluctuations, are dependent on the coherence properties of the wave in the medium at the slab.

Through the use of elementary physical arguments we have deduced the qualitative functional dependence of the spatial phase statistics on optical wave number, propagation distance and the parameters that describe the turbulent medium. We have attempted to delineate the underlying physical mechanisms which produce such fluctuations and as such the derivations presented here should complement the more rigorous analysis presented elsewhere. Finally, the propagation of beam waves in a turbulent medium is determined from a knowledge of the propagation characteristics of spherical waves via the extended Huygens-Fresnel principle [10]. As a result, the qualitative dependence of the characteristics of beam waves (e.g., beam spreading, on axis irradiance) can be obtained directly from the spherical wave coherence length alluded to above [11].

References

1. TATARSKII, V. I., 1971, "Wave Propagation in a Turbulent Medium," McGraw-Hill.
2. TATARSKII, V. I., 1971, "The Effects of the Turbulent Atmosphere on Wave Propagation," National Technical Information Service, U.S. Department of Commerce,Springfield, Va.
3. CHERNOV, L. A., "Wave Propagation in a Random Medium," 1960, McGraw-Hill.
4. BARABANENKOV, Y. N., Y. A. KRAVSTOV, S. M. RYTOV, and V. I. TATARSKII, Sov. Phys.-Usp 13, 551 (1971).
5. PROKHOROV, A. M., F. V. BUNKIN, K. S. GOCHELASVILY, and V. I. SHISHOV, Proc. IEEE 63, 790 (1975).
6. YURA, H. T., J. Opt. Soc. Am. 64, 59 (1974).
7. CLIFFORD, S. F., G. R. OCHS, and R. S. LAURENCE, J. Opt. Soc. Am. 64, 148 (1974).
8. FANTE, R. L., Radio Sci. (New Series) 10, 77 (1975).
9. FANTE, R. L., J. Opt. Soc. Am. 64, 608 (1975).
10. LUTOMIRSKI, R. F., and H. T. YURA, Appl. Opt. 10, 1652 (1971).
11. YURA, H. T., Appl. Opt. 10, 2771 (1971).

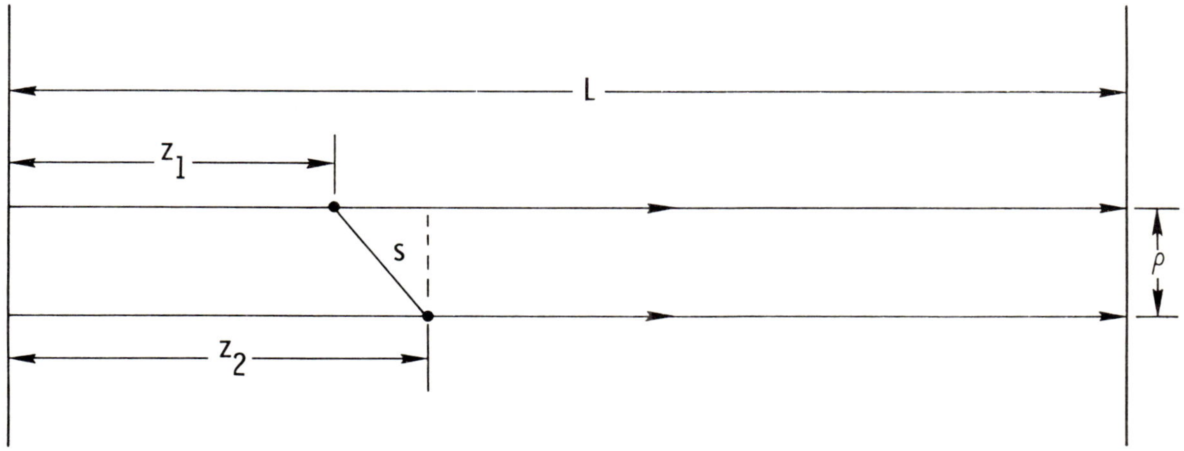

$$s^2 = (z_1 - z_2)^2 + \rho^2$$

Fig. 1. Schematic illustration of the propagation geometry for plane waves

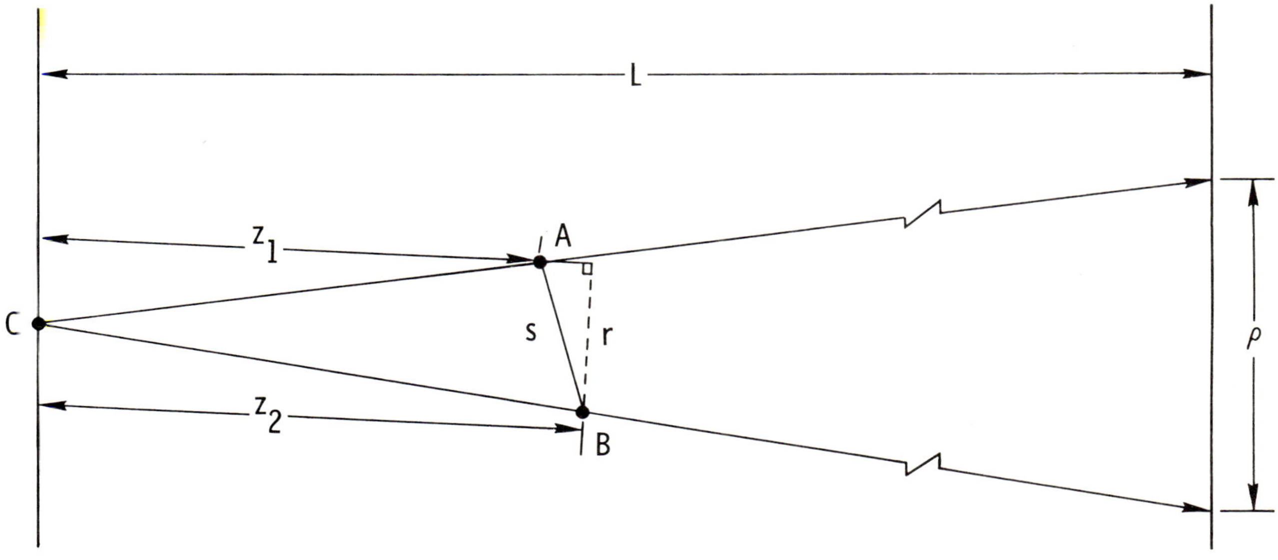

$$\rho << L, \; OA \approx z_1, \; OB \approx z_2$$

$$s^2 = (z_1 - z_2)^2 + r^2 \approx (z_1 - z_2)^2 + \left(\frac{z_1 + z_2}{2L}\right)^2 \rho^2$$

Fig. 2. Schematic illustration of the propagation geometry for spherical waves

THE SEPARATION OF THE OPTICAL TRANSFER FUNCTION IN A TURBULENT MEDIUM

Wayne W. Metheny and Richard B. Philbrick
Optical Sciences Center, University of Arizona
Tucson, Arizona 85721

Abstract

The only condition for separation of the optical transfer function treated in the literature is the long-exposure (ensemble average) condition. Other conditions are considered here. One particular condition based upon the Taylor "frozen atmosphere" hypothesis may be widely applicable for intermediate exposure times. A computer simulation for testing separation in this case is proposed.

Introduction and Analysis

The optical transfer function does not generally separate into two independent factors, although there are a few familiar exceptional cases. By separation in a turbulent medium we mean that the transfer function separates into a factor which depends only on the atmospheric perturbations of the wavefront and a factor which depends only on the optical system. The only case for atmospheric separation treated in the literature is the so-called long exposure condition.[1,2,3,4] The long exposure, or ensemble average condition for a stationary, ergodic random process, is often mistakenly considered as a necessary condition (as well as a sufficient condition) for separation. We consider other conditions in this paper and outline a proposed computer simulation test of one condition based upon the Taylor "frozen atmosphere" hypothesis that seems to have very wide applicability for exposure times much shorter than required for an ensemble average.

The motivation for this study is reflected in the paper by our coworkers R. R. Shannon and W. S. Smith which follows later in this volume. Measurements made on the atmospheric part of the pupil function can be subsequently applied to any optical system since the pupil function always separates. However, the pupil function requires more information for its specification over a finite exposure time than does the transfer function since the latter is integrated over exposure time. Thus, if the transfer function separates, it may be the simpler function to measure.

The transfer function in the general case can be written as follows:

$$\mathrm{OTF}(\eta_x,\eta_y) = \frac{1}{B}\iint_{-\infty}^{\infty}\left[\int_0^T p_a\left(x+\frac{\Delta x}{2},y+\frac{\Delta y}{2};t\right)p_a^*\left(x-\frac{\Delta x}{2},y-\frac{\Delta y}{2};t\right)dt\left|p_o\left(x+\frac{\Delta x}{2},y+\frac{\Delta y}{2}\right)\right|\right.$$
$$\left.\left|p_o\left(x-\frac{\Delta x}{2},y-\frac{\Delta y}{2}\right)\right|e^{i2\pi\Delta w(x,y;\Delta x,\Delta y)}\right]dxdy \quad (1)$$

where the integral is to be evaluated at $\Delta x = \lambda f\eta_x$, $\Delta y = \lambda f\eta_y$.

$p_a(x,y;t)$ = Input wavefront (with atmospheric perturbation) referenced to the exit pupil.

$|p_o(x,y)|$ = Modulus of the optical system pupil function.

$\Delta w(x,y;\Delta x,\Delta y)$= Sheared aberration function.

The integrand (within the brackets) is the sheared pupil function which clearly separates into an atmospheric factor and an optical system factor. When the transfer function separates we write

$$\mathrm{OTF}(\eta_x,\eta_y) = \mathrm{ATF}(\Delta x,\Delta y)\ \mathrm{OTF}_o(\Delta x,\Delta y) \quad (2)$$

where for convenience we leave the terms on the right in terms of the shear dimension, and

$$\mathrm{ATF}(\Delta x,\Delta y) = \frac{\iint_{-\infty}^{\infty}\int_0^T \Delta p_a dt\left|p_o\left(x+\frac{\Delta x}{2},y+\frac{\Delta y}{2}\right)\right|\left|p_o\left(x-\frac{\Delta x}{2}\right)\right|dxdy}{\iint_{-\infty}^{\infty}\left|p_o\left(x+\frac{\Delta x}{2},y+\frac{\Delta y}{2}\right)\right|\left|p_o\left(x-\frac{\Delta x}{2},y-\frac{\Delta y}{2}\right)\right|dxdy} \quad (3)$$

$$\mathrm{OTF}_o(\Delta x,\Delta y) = \frac{1}{B}\iint_{-\infty}^{\infty}\left|p_o\left(x+\frac{\Delta x}{2},y+\frac{\Delta y}{2}\right)\right|\left|p_o\left(x-\frac{\Delta x}{2},y-\frac{\Delta y}{2}\right)\right|e^{i2\pi\Delta w(x,y;\Delta x,\Delta y)}dxdy \quad (4)$$

where Δp_a = Sheared atmospheric pupil function

The modulus of the optical system pupil function appears in the ATF, but this is usually just a binary function which determines the region over which the integration is taken (i.e. the "overlap" region).

The most familiar case for separation of the transfer function in image formation theory is the separation due to weak aberrations. That is, if the aberrations are weak, the effect on the transfer function is the same (a function of the RMS wavefront deviation) regardless of the source of the aberration. This can be extended to cover weak turbulence,[5] and applies to RMS wavefront deviations of less than about 0.15 waves. There is a separation in very strong turbulence as well if the sheared wavefront perturbations are rapidly varying over the aperture and the sheared aberration function is slowly varying. However, the strong turbulence condition is probably of less interest since the resulting image is poor in any event.

Another familiar exceptional case in the theory of image formation is the separation due to image motion. The image motion usually considered in the theory occurs at a uniform velocity over the exposure simplifying the analysis, but separation of the transfer function is not limited to that restriction. The effective psf for the exposure is the optical system psf or impulse response convolved with the time history of the image motion; separation of the transfer function into a factor which depends only on the optical system and a factor which depends only on the image motion follows by the convolution theorem. Image motion produced by turbulence is called image dance or agitation and is not uniform in velocity. It is caused by large scale turbulence, i.e. turbulent cells large with respect to the aperture. Motion of the image is equivalent to a tilt of the pupil function or wavefront. Wavefront tilt due to turbulence has been studied using some theoretical arguments.[3,6] A significant part of the theoretical long exposure transfer function (i.e. ensemble average) is often caused by wavefront tilt.

A last familiar case for separation applies to the long exposure transfer function in a turbulent medium with time varying wavefronts. This separation is actually a special case of a more general condition which can be applied in a different way with the aid of Taylor's hypothesis. With time varying wavefronts the effective transfer function is proportional to the sheared pupil function integrated over the aperture and then integrated over the exposure time. Formally, however, the order of integration may be interchanged so that the transfer function is proportional to the sheared pupil function at one point in the aperture, integrated over time and then integrated over the aperture. The atmospheric part of the transfer function is separable if the time integration for one point in the aperture is independent of the position in the aperture. It follows that if each point of the aperture sees the same instantaneous values of the sheared pupil function (in whatever order) over the exposure time, then the transfer function separates. One way to insure (formally) that each point of the aperture sees the same values is to impose the condition that each point sees all possible members of the statistical ensemble which describes the current state of turbulence; i.e. each point sees the ensemble average value of the sheared pupil function, assuming that an ergodic, stationary random process is a valid representation of the turbulence. This is a statement of the long exposure separation condition. The ensemble average condition, while sufficient for separation, is often mistakenly treated in the literature as if it were a necessary condition.

Separation due to image motion or wavefront tilt can be interpreted as a trivial example of the more general condition for separation. Wavefront tilt produces a contribution to the phase of the sheared pupil function which is constant over the whole aperture. Thus each point of the aperture sees the same values of the sheared pupil function due to wavefront tilt (in this case in exactly the same order) over the exposure time for any exposure time. It is tempting (and probably correct) to speculate that the large scales of turbulence which cause wavefront tilt take the longest time to reach an ensemble average value of the sheared pupil function; thus, if true, effective separation of the transfer function would occur much quicker than the ensemble average condition implies.

Separation in one dimension for all scales of turbulence occurs if Taylor's hypothesis is applied to describe to the time varying wavefronts; then the sheared pupil function is "frozen" and is swept across the pupil by the wind. Each point in the aperture then sees the same values of the sheared pupil function as any other point with the same y coordinates (wind in the x-direction) except for the end points. Thus, if the exposure time is long enough to reduce the effect of the end points, the transfer function for a camera with a strip aperture in the x-direction clearly separates.

Formally, the time integral contained in the OTF expressions becomes (with a frozen atmosphere):

$$\int_0^T p_a\left(x - vt + \frac{\Delta x}{2}, y + \frac{\Delta y}{2}\right) p_a\left(x - vt - \frac{\Delta x}{2}, y - \frac{\Delta y}{2}\right) dt \qquad (5)$$

which for simplicity can be written

$$\int_0^T g(x - vt, y)\, dt \qquad (6)$$

or, rewriting

$$\int_0^T g'\left(\frac{x}{v} - t, y\right) dt \qquad (7)$$

if $x/v << T$, then the above integral is approximately independent of x:

$$\int_0^T g'(\frac{x}{v} - t,y)dt \cong \int_0^T g'(t,y)dt \quad (\text{small } \frac{x}{vT}) \tag{8}$$

Then the OTF separates in the x-direction, but possible y-variations remain.

$$OTF = \frac{1}{B}\int_{y_1}^{y_2}\left[\int_0^T g'(t,y)dt \int_{x_1}^{x_2} e^{i2\pi\Delta w(x,y;\Delta x)}dx\right]dy \tag{9}$$

where the limits of integration specify the appropriate overlap region (a function of Δx).

We contend that the y integrand (above) is a weak function of y and that hence the partial separation is, for practical purposes, a complete separation over x and y. This is partly true because the aberration function itself, when integrated in the indicated manner over the x-direction, may become a fairly weak function of y (and hence give y separation); this has been demonstrated for some small 3rd order aberrations.(5) The atmospheric part integrated over x (i.e. or over time) is probably an even weaker function of y. That is, if the exposure time is long enough for three aperture lengths of the atmosphere to be swept by, then the ATF is dominated by the x or t variation and, practically, the y integration acts only as a weighting function. These two smoothed functions together reinforce the contention.

Computer Simulation

In this section we present a proposal for a computer simulation to test the conslusions of the previous section. The test would require a modification of the 'OTF' and 'FROLIC' programs developed at the Optical Sciences Center, University of Arizona, by James Rancourt and Barbara Fell.(7) 'OTF' calculates the optical transfer function (OTF) and point spread function (PSF) for a user-specified exit pupil phase distribution. 'FROLIC' traces real rays through a lens system and includes 'OTF' as an overlay to make OTF and PSF calculations of the computed exit pupil phase error.

The simulation would test separation of the atmospheric ATF and the lens OTF_o through parallel calculations. The OTF of the atmosphere and lens taken as one optical system would be computed first using the separation hypothesis and then directly by adding the lens and atmospheric phase perturbations. The results of the two computations could then be compared.

The first step in the simulation is to synthesize the phase wavefront seen by an aperture moving rapidly in what we will call the 'x' direction. If the aperture is moving sufficiently fast a frozen atmosphere can be assumed. Thus one large wavefront of light propagating from the point of interest could be created and the aperture swept across it. This aperture will be assumed to be slightly larger than thirty centimeters in diameter so that it can be represented by a 31x31 complex matrix. The simulated wavefront would be a strip of 31 points in the 'y' direction by 128 points in the 'x' direction allowing the aperture to move about three diameters.

A random wavefront can be created from a theoretical phase power spectrum such as that given by Tatarski.(8) The method would involve synthesis of a two-dimensional Fourier transformation of a possible wavefront from the power spectrum. Inverse transformation then would yield the wavefront.

In practice, two wavefronts would be created and summed to give a wider range of frequencies. One wavefront would be 64x64 points (to fit available fast Fourier transform routines) with a scale of 4cm/point (i.e. 2.56x2.56 meters). Frequencies between .39 and 6.25 cycles/meter would be present on this matrix. An 8x32 point section of this matrix would be interpolated to give 32x128 points or 1cm/point.

A second matrix of 64x64 with a scale of 1cm/point would be used to simulate higher frequencies between 7.81 and 50.00 cycles/meter. A strip of 31x64 points would be selected from this matrix and replicated once in the 'x' direction to create a strip of 31x128cm.

These two strips would then be added to give a 31x128cm wavefront with frequencies between .39 and 50.0 cycles/meter. Unreplicated phase variation wavelengths would range from half the aperture diameter to twice the length of the exposure strip.

The next step is to select a suitable lens system with an aperture just larger than thirty centimeters in diameter. The quality of this system must be sufficiently good so that the phase perturbation at the exit pupil plus the atmospheric phase variation never gives a point to point (i.e. over 1cm) phase difference greater than a quarter wave (note that this is a limitation on the wavefront variation as well). Given such a lens the phase variation at the exit pupil and the OTF_o (of the system) can be calculated using the 'FROLIC' program modified to save the exit pupil phase variation and OTF_o on disc or tape.

Now the OTF_S or OTF of lens plus atmosphere assuming separation can be calculated. The PSF of the aperture with atmospheric phase perturbation alone is calculated many times as the aperture is stepped over the wavefront strip. These individual PSF's for atmospheric effect are averaged and an OTF_A is calculated for the integrated PSF. A diffraction limited OTF_D for the aperture would be calculated as well to avoid squaring the effect of the aperture and the ATF is OTF_A divided by OTF_D.

Then: $$OTF_s = OTF_o \times \left|\frac{OTF_A}{OTF_D}\right| = OTF_o \times ATF \tag{10}$$

The direct calculation of the OTF of atmosphere and lens (OTF) is straight forward also. The atmospheric wavefront and exit pupil phase variations are added and PSF's calculated as the aperture is stepped along the wavefront strip. The PSF's are integrated and OTF calculated directly.

These matrices can then be compared for various separations in the aperture in 'x', 'y' and 45° directions. The results would be for three aperture diameters of movement but intermediate results could be given for exposure sweeps of 0, 1/2, 1 and 2 aperture diameters.

Although it would not cover all possibilities, this procedure would test the separation theory in its most questionable area -- that of phase perturbations with wavelengths approximately the same as the aperture diameter. The simulation could be extended, with considerable time and money, to include:

1. Variation of level of turbulence.
2. Variation of the phase spectrum slope used to synthesize the data.
3. Inclusion of scintillation.
4. Larger matrix for longer wavelength phase variation.
5. Larger matrix for a smaller scale to reduce the effect of the quarter wave step restriction.

References

1. O'Neill, Edward L., Introduction to Statistical Optics, Addison-Wesley, Reading, Mass., 1963.
2. Hufnagel, R. E., and Stanley, N. R., J. Opt. Soc. Am., 54, 1, 52-61, (1964).
3. Fried, D. L., J. Opt. Soc. Am., 56, 10, 1372-1379, (1966).
4. Shack, R. V., Opt. Sci. Center Tech. Rept. #19, U. of Az., (1967).
5. Metheny, W. W., unpublished.
6. Yura, H. T., J. Opt. Soc. Am., 63, 5, 567-572.
7. Hooker, B.; Shack, R. V. et al, "Study and Investigation in Optical Technology", AFAL-TR-72-152.
8. Tatarski, V. I., Wave Propagation in a Turbulent Medium, McGraw-Hill, N.Y., 1961.

VARIETIES OF ISOPLANATISM*

David L. Fried
Optical Science Consultants
P. O. Box 388, Yorba Linda, California 92686

Abstract

Isoplanatism is a somewhat exotic term used to indicate that the transfer function of an optical system is dependent on the field-angle, or to denote the region, called the isoplanatic patch, over which the transfer function is virtually independent of field-angle. For many years, the term has been used mostly as a sort of "charm" to "ward-off" the possibility that others might think we were unaware of mathematical subtleties we intended to ignore — because they were physically inconsequential. However, isoplanatism assumes physical significance in imaging through turbulence. The need to take account of this has created some confusion since different types of imagery have different isoplanatic dependencies and so should be denoted by different terms. There are a variety of different effects each of which can be classified as isoplanatism — each distinct in its dependence on the propagation path. We have identified five distinct varieties of isoplanatism, which we call 1) predetection compensation isoplanatism, 2) post-detection short-exposure imagery compensation isoplanatism, 3) post-detection long-exposure compensation isoplanatism, 4) angle-of-arrival isoplanatism, and 5) speckle interferometry isoplanatism. Formulas giverning each of these types of isoplanatism will be presented.

Introduction

The term "isoplanatism" is used to define a dependence, or rather a lack of dependence of the performance of an optical system on field angle. It had its origin in the need, in defining the OTF of a lens, to assume that the transfer function of the lens was invariant with position in the focal plane so that the OTF concept would be meaningful. Recognizing that the transfer function was, of course, not really stationary, the concept of the isoplanatic patch was defined. It denotes a region in the focal plane, or in the field-of-view, for which the lens' transfer function was sufficiently close to stationary that it could be considered to be exactly stationary. The lens OTF could then be defined over that region by assuming that the transfer function in that region actually applied, for mathematical (i.e., Fourier transform) purposes, over an infinite extent.

As a practical matter, all this is quite reasonable, and gives meaningful and useful results — but results that could have been obtained directly, without ever considering isoplanatism. In a formal mathematical sense, however, isoplanatism is just a way of acknowledging the existence of a mathematical difficulty, with the intent of passing it by — without resolving it. In view of this, the matter of isoplanatism has for many years languished as an obscure comment or footnote in various works on the OTF of imaging systems.[1]

With the development of interest in various rather sophisticated concepts of imaging through the atmosphere, the subject of isoplanatism once again appears, only this time with more practical significance. In imaging through atmospheric turbulence, the term "isoplanatism" is used to denote that a particular effect is independent of field angle. For some concepts, this independence makes the difference between the concept's working or not working, and the isoplanatic patch is a quantity with a very practical meaning — it denotes the largest size field-of-view for which the concept will work, or rather a parameter which governs how rapidly the performance of a concept falls off with field-of-view size.

Inasmuch as there are many different concepts relating to the problem of coping with the effects of atmospheric turbulence on imaging systems, it is reasonable to expect differences in the effects associated with field angle dependence, i.e., different isoplanatism dependencies — hence the title of the paper.

In the following sections, we shall develop the basic theory governing these different types of isoplanatism. We shall consider first isoplanatism for predetection compensation imaging,[2] and after that we shall treat isoplanatism for angle-of-arrival fluctuations. We shall then consider isoplanatism as it impacts post-detection compensation of imagery,[3] treating both long- and short-exposures. Finally, we shall present a formulation for the isoplanatic dependence of speckle interferometry.[4]

It will be noted that the isoplanatic dependence for post-detection compensation of long-exposure imagery is fundamentally different and simpler than the dependence of the other four processes. It will further be noted that the isoplanatism dependencies of these other four processes have an inner similarity, though by no means an equivalence. Only the isoplanatic dependence of post-detection compensation of short-exposure imagery and of speckle interferometry have a one-for-one correspondence — between

* This work was sponsored by the U. S. Air Force Rome Air Development Center, with funds provided by the Defense Advanced Research Projects Agency under Contract No. F30602-74-C-0115 and Contract No. F30602-76-C-0005 The RADC Project Engineer is Mr. Don Hanson.

compensation error for the former, and signal strength for the latter.[5] With the exception of this one pairing, we shall conclude that the isoplanatic dependencies of the different processes are of the same approximate scale, but that significant differences in the exact dependence can exist.

Predetection Compensation

Predetection compensation is a process by which an adaptive optics imaging system formulates an estimate of the wavefront distortion caused by atmospheric turbulence and produces a compensating distortion of the optics in the entrance aperture, so that a nominally diffraction-limited image can be formed. In its simplest form, a reference point source is present in the field-of-view. It would be used as a basis for determining the required wavefront distortion correction for imaging of some other small object also in the field-of-view. Though there are other ways of defining a reference to be used for formulation of an estimate of the required wavefront distortion compensation, this approach contains the fundamental isoplanatism dependence of the other approaches, and we shall base our analysis on it.

We define the isoplanatism dependence of predetection compensation imaging as the dependence of the OTF of the image for image frequency $\vec{f}$, upon the angular separation, $\vec{\vartheta}$, between the reference point source and the object to be imaged. If the reference is at location $\vec{\theta}_1$, and the object to be imaged, which we shall take to be a point source, is at location $\vec{\theta}_2$, then we can write for the wavefront error due to isoplanatism dependence after compensation, at location $\vec{r}$ in the entrance aperture

$$\Delta\phi(\vec{r},\vec{\vartheta}) = \phi(\vec{r},\vec{\theta}_2) - \phi(\vec{r},\vec{\theta}_1) \quad , \tag{1}$$

where

$$\vec{\vartheta} = \vec{\theta}_2 - \vec{\theta}_1 \quad , \tag{2}$$

and $\phi(\vec{r},\vec{\theta})$ is the turbulence-induced phase shift for propagation from a point source at $\vec{\theta}$ to aperture position $\vec{r}$. (Here and throughout this analysis, we shall suppress the relatively minor log-amplitude fluctuations.) We can write for the predetection compensated wavefront in the aperture plane due to the source at $\vec{\theta}_2$,

$$U(\vec{r}) = W(\vec{r}) \exp\left[i\Delta\phi(\vec{r},\vec{\vartheta})\right] \quad , \tag{3}$$

where $W(\vec{r})$ is a function defining a circular aperture of diameter D, according to the equation

$$W(\vec{r}) = \begin{cases} 1 & \text{if} \quad |\vec{r}| \le \tfrac{1}{2}D \\ 0 & \text{if} \quad |\vec{r}| > \tfrac{1}{2}D \end{cases} \tag{4}$$

It is easy to show that the amplitude in the image plane at position $\vec{y}$ can be written as

$$u(\vec{y}) = A \int d\vec{x}\, U(\vec{r}) \exp(-ik\vec{r}\cdot\vec{y}) \quad , \tag{5}$$

where A is a constant of proportionality and $k = 2\pi/\lambda$ is the optical wave number. The instantaneous OTF of the system for the imaging of a source at $\vec{\theta}_2$ can be calculated as the Fourier transform of the focal plane intensity associated with the amplitude $u(\vec{y})$, since the source we have treated to calculate $u(\vec{r})$ is a point source. Thus we can write for the OTF at image frequency $\vec{f}$

$$\tau(\vec{f}) = B \int d\vec{y}\, |u(\vec{y})|^2 \exp(2\pi i \vec{f}\cdot\vec{y}) \quad , \tag{6}$$

where B is a constant of proportionality chosen to make $\tau(0)$ equal to unity.

If we substitute Eq. (5) into Eq. (6) twice, make the produce of integrals into a multiple integral and then suitably reduce the integral by identification of delta functions, we get

$$\tau(\vec{f}) = A^2 B \int d\vec{x}\, U^*(\vec{r}-\lambda\vec{f})\, U(\vec{r}) \quad . \tag{7}$$

If we substitute Eq. (3) into Eq. (7), we get

$$\tau(\vec{f}) = A^2 B \int d\vec{r}\, W(\vec{r}-\lambda\vec{f})\, W(\vec{r}) \exp\{i[\Delta\phi(\vec{r},\vec{\vartheta}) - \Delta\phi(\vec{r}-\lambda\vec{f},\vec{\vartheta})]\} \quad . \tag{8}$$

From Eq. (8), it follows that the ensemble average OTF for predetection compensation imaging can be written as

$$\langle \tau(\vec{f}) \rangle = \tau_0(\vec{f})\, M(\lambda\vec{f},\vec{\vartheta}) \quad , \tag{9}$$

where $\tau_0(\vec{f})$ is the diffraction-limited OTF, given by

$$\tau_0(\vec{f}) = A^2 B \int d\vec{r}\, W(\vec{r}-\lambda\vec{f})\, W(\vec{r}) \quad , \tag{10}$$

and

$$M(\lambda\vec{f},\vec{\vartheta}) = \langle \exp\{i[\Delta\phi(\vec{r},\vec{\vartheta}) - \Delta\phi(\vec{r}-\lambda\vec{f},\vec{\vartheta})]\} \rangle \quad . \tag{11}$$

It is particularly significant in developing Eq. (9) from Eq. (8) to note that by virtue of the statistical stationarity of propagation effects, $M(\lambda\vec{f},\vec{\vartheta})$ as defined in Eq. (11), though seemingly having an

$\vec{r}$-dependence, actually is independent of $\vec{r}$. We identify $M(\lambda\vec{f},\vec{\vartheta})$ as defining the isoplanatic dependence of predetection compensation imaging. For $\vec{\vartheta} \equiv 0$, M obviously has unity value, and we may suppose for $\vec{\vartheta}$ very large, M has a very small value. The exact nature of the $\vec{f}$- and $\vec{\vartheta}$-dependencies will be the subject of our consideration in the balance of this section.

By virtue of the fact that $\phi(\vec{r},\vec{\theta})$ is a gaussian random variable, it follows that Eq. (11) can be re-written in the form

$$M(\lambda\vec{f},\vec{\vartheta}) = \exp\left[-\mathcal{M}(\lambda\vec{f},\vec{\vartheta})\right] \quad , \tag{12}$$

where

$$\mathcal{M}(\lambda\vec{f},\vec{\vartheta}) = \tfrac{1}{2}\langle[\Delta\phi(\vec{r},\vec{\vartheta}) - \Delta\phi(\vec{r}-\lambda\vec{f},\vec{\vartheta})]^2\rangle \quad . \tag{13}$$

If we substitute Eq. (1) into Eq. (13), and define $\mathcal{D}_{I,\phi}(\vec{\rho},\vec{\vartheta})$ by the equation

$$\mathcal{D}_{I,\phi}(\vec{\rho},\vec{\vartheta}) = \langle[\phi(\vec{r}+\vec{\rho},\vec{\theta}+\vec{\vartheta}) - \phi(\vec{r},\vec{\theta})]^2\rangle \quad , \tag{14}$$

then we can write

$$\mathcal{M}(\lambda\vec{f},\vec{\vartheta}) = \mathcal{D}_{I,\phi}(0,\vec{\vartheta}) + \mathcal{D}_{I,\phi}(\lambda\vec{f},0) - \tfrac{1}{2}\mathcal{D}_{I,\phi}(\lambda\vec{f},\vec{\vartheta}) - \tfrac{1}{2}\mathcal{D}_{I,\phi}(\lambda\vec{f},-\vec{\vartheta}) \quad . \tag{15}$$

It can be shown[6] that for a point source well above the atmosphere

$$\mathcal{D}_{I,\phi}(\vec{\rho},\vec{\vartheta}) = 2.91\, k^2 \int_{\text{Path}} ds\, C_N^2\, |\vec{\rho}+\vec{\vartheta}s|^{5/3} \quad , \tag{16}$$

where the path of integration runs from $s = 0$ at the ground up to the source.

If we substitute Eq. (16) into Eq. (15) and appropriately combine and simplify results, we get

$$\mathcal{M}(\lambda\vec{f},\vec{\vartheta}) = 2.91\, k^2 \int_{\text{Path}} ds\, C_N^2\, \{(\vartheta s)^{5/3} + (\lambda f)^{5/3} + \tfrac{1}{2}[(\lambda f)^2 + 2(\lambda f)(\vartheta s)\cos\phi + (\vartheta s)^2]^{5/6}$$
$$- \tfrac{1}{2}[(\lambda f)^2 - 2(\lambda f)(\vartheta s)\cos\phi + (\vartheta s)^2]^{5/6}\} \quad , \tag{17}$$

where ϕ is the orientation angle between the directions defined by the vectors $\vec{\vartheta}$ and $\vec{f}$. We may consider the combination of Eq.'s (17) and (12) [or Eq. (17) alone] as a definition of predetection compensation isoplanatism. Further simplification is a straightforward matter, but depends on the particular distribution of C_N^2-values, as well as the $\vec{\vartheta}$- and $\vec{f}$-values of interest. This is as far as we shall go with this matter, though we have pursued the problem of simplification elsewhere.[7] We next take up the matter of angle-of-arrival isoplanatism.

Angle-of-Arrival

The turbulence-induced random wavefront distortion may, in certain cases, be considered to be principally a matter of random wavefront tilt.[8] In such a case, most of the effect of the turbulence can be compensated by suitable tilt compensation — but in certain cases we have to use a reference not exactly co-directional with the direction of interest to define the required tilt correction. As an example of this, consider a ground-based laser transmitter sending a beam to a satellite carrying a beacon. The turbulence-induced fluctuations in the apparent position of the beacon would be tracked and compensated in the laser transmission, using a shared aperture for transmission and beacon reception. Because of the round trip transit time of the light from the satellite to the ground and back, and the motion of the satellite during that time, it would be necessary to utilize a point-ahead bias between the beacon direction and the transmission direction.

As far as the atmospheric turbulence isoplanatism effects are concerned, we are dealing with two different directions, $\vec{\theta}_1$, the direction to the beacon, and $\vec{\theta}_2$, the direction of transmission, with angular separation $\vec{\vartheta}$. By virtue of reciprocity considerations,[9] we know that the proper transmitter tilt is actually the wavefront tilt we would see if we could view a beacon at $\vec{\theta}_2$. We are interested in knowing how much our angle-of-arrival information based on the beacon at $\vec{\theta}_1$ differs from the information we would obtain from a beacon at $\vec{\theta}_2$, as this is a measure of the failure of tilt compensation in the laser transmitter. We shall take as a measure of this the mean square difference in the two angles and define this as the angle-of-arrival isoplanatism dependence. We write this as

$$\mathcal{D}_{I,\alpha}(\vec{\vartheta}) = \langle[\vec{\alpha}(\vec{\theta}_1) - \vec{\alpha}(\vec{\theta}_2)] \cdot [\vec{\alpha}(\vec{\theta}_1) - \vec{\alpha}(\vec{\theta}_2)]\rangle \quad , \tag{18}$$

where $\vec{\alpha}(\vec{\theta})$ is a vector quantity denoting the two components of the angle-of-arrival as seen in an aperture of diameter D, viewing a source at $\vec{\theta}$.

We start our calculations by noting that, as we have shown previously,[10]

$$\vec{\alpha}(\vec{\theta}) = \frac{32\lambda}{\pi^2 D^4} \int d\vec{r}\; W(\vec{r})\, \vec{r}\, \phi\; (\vec{r},\vec{\theta}) \qquad . \tag{19}$$

Substituting Eq. (19) into Eq. (18), making a double integral out of a produce of integrals, and interchanging the order of integration and ensemble averaging, we get

$$\mathcal{D}_{I,\alpha}(\vec{\vartheta}) = \left(\frac{32\lambda}{\pi^2 D^4}\right)^2 \iint d\vec{r}\, d\vec{r}\,'\, W(\vec{r})\, W(\vec{r}\,')\, \vec{r}\cdot\vec{r}\,'[\tfrac{1}{2}\mathcal{D}_{I,\phi}(\vec{r}-\vec{r}\,',\vec{\vartheta}) + \tfrac{1}{2}\mathcal{D}_{I,\phi}(\vec{r}-\vec{r}\,',-\vec{\vartheta}) - \mathcal{D}_{I,\phi}(\vec{r}-\vec{r}\,',0)] . \tag{20}$$

In obtaining Eq. (20), we have made use of the fact that in accordance with Eq. (14), we can show by suitable algebraic manipulation that

$$\langle[\phi(\vec{r},\vec{\theta}_1)-\phi(\vec{r},\vec{\theta}_2)][\phi(\vec{r}\,',\vec{\theta}_1)-\phi(\vec{r}\,',\vec{\theta}_2)]\rangle = \tfrac{1}{2}\mathcal{D}_{I,\phi}(\vec{r}-\vec{r}\,',\vec{\vartheta}) + \tfrac{1}{2}\mathcal{D}_{I,\phi}(\vec{r}-\vec{r}\,',-\vec{\vartheta}) - \mathcal{D}_{I,\phi}(\vec{r}-\vec{r}\,',0) \quad . \tag{21}$$

Now, if we make use of Eq. (16) and of the fact that

$$\iint d\vec{r}\, d\vec{r}\,'\, W(\vec{r})\, W(\vec{r}\,')\, \vec{r}\cdot\vec{r}\,' \equiv 0 \qquad , \tag{22}$$

then Eq. (20) can be cast in the form

$$\mathcal{D}_{I,\alpha}(\vec{\vartheta}) = 2.91 \left(\frac{64}{\pi D^2}\right)^2 \int_{\text{Path}} ds\, C_N^2 \iint d\vec{r}\, d\vec{r}\,'\, W(\vec{r})\, W(\vec{r}\,')\, \vec{r}\cdot\vec{r}\,'[\tfrac{1}{2}|\vec{r}-\vec{r}\,'+\vec{\vartheta}s|^{5/3}$$
$$+ \tfrac{1}{2}|\vec{r}-\vec{r}\,'-\vec{\vartheta}s|^{5/3} - |\vec{r}-\vec{r}\,'|^{5/3} - |\vec{\vartheta}s|^{5/3} \qquad . \tag{23}$$

By introducing the variables

$$\vec{u} = (\vec{r}-\vec{r}\,')/D \qquad , \tag{24a}$$

$$\vec{v} = \tfrac{1}{2}(\vec{r}+\vec{r}\,')/D \qquad , \tag{24b}$$

and defining

$$\tilde{W}(\vec{x}) = \begin{cases} 1 & \text{if} \quad |\vec{x}| \le \tfrac{1}{2} \\ 0 & \text{if} \quad |\vec{x}| > \tfrac{1}{2} \end{cases} \qquad , \tag{25}$$

we can, with a change of variables of integration, cast Eq. (23) in the form

$$\mathcal{D}_{I,\alpha}(\vec{\vartheta}) = 2.91\,(64/\pi)^2\, D^{-1/3} \int_{\text{Path}} ds\, C_N^2 \iint d\vec{u}\, d\vec{v}\, \tilde{W}(\vec{v}+\tfrac{1}{2}\vec{u})\, \tilde{W}(\vec{v}-\tfrac{1}{2}\vec{u})(v^2 - \tfrac{1}{4}u^2)[\tfrac{1}{2}|\vec{u}+\vec{\vartheta}s/D|^{5/3}$$
$$+ \tfrac{1}{2}|\vec{u}-\vec{\vartheta}s/D|^{5/3} - u^{5/3} - (\vartheta s/D)^{5/3}] \qquad . \tag{26}$$

To simplify Eq. (26), we note that the $\vec{v}$-integration can be performed using standard indefinite integrals, yielding the result that

$$\int d\vec{v}\, \tilde{W}(\vec{v}+\tfrac{1}{2}\vec{u})\, W(\vec{v}-\tfrac{1}{2}\vec{u})(v^2 - \tfrac{1}{4}u^2) = \frac{1}{16}[\cos^{-1}(u) - (3u-2u^2)(1-u^2)^{1/2}] \qquad . \tag{27}$$

If we substitute Eq. (27) into Eq. (26), the result can be cast in the form

$$\mathcal{D}_{I,\alpha}(\vec{\vartheta}) = 2.91\,(16/\pi)^2\, D^{-1/3} \int_{\text{Path}} ds\, C_N^2 \int_0^{2\pi} d\phi \int_0^1 u\, du\, [\cos^{-1}(u) - (3u-2u^2)(1-u^2)^{1/2}$$
$$\times \{\tfrac{1}{2}[u^2 + 2u\,(\vartheta s/D)\cos\phi + (\vartheta s/D)^2]^{5/6} + \tfrac{1}{2}[u^2 - 2u\,(\vartheta s/D)\cos\phi + (\vartheta s/D)^2]^{5/6}$$
$$- u^{5/3} - (\vartheta s/D)^{5/3}\} \qquad . \tag{28}$$

This result can not be reduced any further by analytic techniques, though for computational purposes it is convenient to treat the ϕ- and u-integrations as defining a function of the variable $\vartheta s/D$ for use with various C_N^2 distributions. For our purposes, we may consider Eq. (28) as our final result defining angle-of-arrival isoplanatism. It is interesting to compare Eq. (28) with Eq. (17), the corresponding result for predetection compensation isoplanatism. The similarities and the differences are quite obvious.

With these results in hand, we shall now turn our attention to the isoplanatism dependencies of post-detection compensation. We take this up in the next two sections.

Post-Detection Compensation — Long-Exposure

Post-detection processing relies on the formulation of an estimate of what the OTF of the imaging process was and compensation of the image by Fourier transforming the image, dividing by the estimated OTF, and then inverse Fourier transforming to recover a compensated image. The basic limitation to this process is due to noise in the detected image. The compensation process amplifies this noise — the smaller the OTF, the greater the amplification. As a consequence, it is necessary to truncate the compensation at some image frequency so as to avoid attempting to compensate frequencies for which the OTF of the detection process is too small.

A secondary and much more minor consideration for post-detection compensation of a long-exposure image is the accuracy of our knowledge of the OTF of the detection process. If this is in error, the post-detection compensated image will have a non-unity effective OTF. We expect this effect to be very minor, but its consideration provides a convenient lead into the same problem for short-exposure imagery, where the matter is more significant, and which we will treat in the next section.

The OTF for long-exposure astronomical imagery is well known to have the form[11]

$$\tau_{LE}(f;\theta) = \tau_0(f) \exp\left[-3.44\,(\lambda f/r_0)^{5/3} \sec\theta\right] \qquad (29)$$

where θ is the zenith angle, and r_0 is a parameter calculated for the vertical propagation path. In the most straightforward approach to post-detection compensation of a long-exposure image, a point source near the object of interest would also be imaged. From that image, we would determine the OTF by Fourier transforming and use that to compensate the image of the object of interest. Our concern is with the discrepancy in this compensation if the object of interest is at the zenith angle θ_1 and the reference point source is at zenith angle θ_2 .

We can write the compensation discrepancy as

$$\Delta_{LE}(f, \theta_1 - \theta_2) = \left[\tau_{LE}(f;\theta_1) - \tau_{LE}(f;\theta_2)\right]^2 \quad . \qquad (30)$$

Making use of Eq. (29), we can show that

$$\tau_{LE}(f;\theta_2) = \tau_{LE}(f;\theta_1) \exp\left[-3.44\,(\lambda f/r_0)^{5/3} \sec\theta \tan\theta\,(\theta_2 - \theta_1)\right] \quad . \qquad (31)$$

Taking account of the fact that $\theta_2 - \theta_1$ will in general be a small number, and that because of image noise considerations we would never attempt post-detection compensation for values of f that made $3.44\,(\lambda f/r_0)^{5/3} \sec\theta$ very large, it follows that the exponential in Eq. (31) can be expanded in a two-term power series. Thus we get

$$\tau_{LE}(f;\theta_2) = \tau_{LE}(f;\theta_1)\left[1 + 3.44\,(\lambda f/r_0)^{5/3} \sec\theta \tan\theta\,(\theta_1 - \theta_2)\right] \quad . \qquad (32)$$

If we substitute Eq. (32) into Eq. (30) and take account of the smallness of the second quantity in the square brackets in Eq. (32), we obtain the result that

$$\Delta_{LE}(f, \theta_1 - \theta_2) = \left[\tau_{LE}(f;\theta_1)\right]^2 \left[3.44\,(\lambda f/r_0)^{5/3} \sec\theta \tan\theta\,(\theta_1 - \theta_2)\right]^2 \quad . \qquad (33)$$

Dropping the explicit θ-dependence in τ_{LE} and letting

$$\delta\theta = \theta_2 - \theta_1 \quad , \qquad (34)$$

it is convenient to rewrite Eq. (33) in the form

$$\Delta_{LE}(f, \delta\theta) = \left\{\tau_{LE}(f)\,\ln\left[\tau_{LE}(f)\right] \tan\theta\;\delta\theta\right\}^2 \quad , \qquad (35)$$

which will serve as our final result for post-detection compensation isoplanatism. In essentially every case of interest, Δ_{LE} is very small.

Post-Detection Compensation — Short-Exposure

In the case of short-exposure imagery each image is, in a sense, a random result. While in long-exposure imagery the OTF is a deterministic, non-random quantity, in short-exposure imagery the OTF for each image is randomly different from the OTF for any other exposure. Only if the two exposures are made simultaneously and viewing in very nearly the same direction is there much correlation between the two OTF's.

If we are imaging an object located in the direction $\vec{\theta}_1$, and simultaneously imaging a reference point source located in the nearby direction $\vec{\theta}_2$, then we would achieve a post-detection compensated image OTF which would be the ratio of the two short-exposure OTF's, $\tau_{SE}(\vec{f},\vec{\theta}_1)$ and $\tau_{SE}(\vec{f},\vec{\theta}_2)$. The discrepancy in this compensation would be

$$\Delta_{SE}(\vec{f},\vec{\vartheta}) = |\tau_{SE}(\vec{f};\vec{\theta}_1) - \tau_{SE}(\vec{f};\vec{\theta}_2)|^2 \quad . \tag{36}$$

In Eq. (36), unlike Eq. (30), we have explicitly used the absolute value notation since τ_{SE} is, in general, a complex quantity. Inasmuch as $\tau_{SE}(\vec{f},\vec{\theta})$ is a random variable, then so is $\Delta_{SE}(\vec{f},\vec{\vartheta})$. What we should concern ourselves with is the mean square value of this quantity $\langle \Delta_{SE}(\vec{f},\vec{\vartheta})\rangle$, as this is a measure of the mean square error in our compensation due to isoplanatism considerations.

It is convenient to write for the quantity of interest

$$\begin{aligned}\langle \Delta_{SE}(\vec{f},\vec{\vartheta})\rangle &= \langle |\tau_{SE}(\vec{f};\vec{\theta}_1) - \tau_{SE}(\vec{f};\vec{\theta}_2)|^2\rangle \\ &= \langle |\tau_{SE}(\vec{f};\vec{\theta}_1)|^2\rangle + \langle |\tau_{SE}(\vec{f};\vec{\theta}_2)|^2\rangle \\ &\quad - \langle \tau_{SE}(\vec{f};\vec{\theta}_1)\,\tau_{SE}^*(\vec{f};\vec{\theta}_2)\rangle - \langle \tau_{SE}^*(\vec{f};\vec{\theta}_1)\,\tau_{SE}(\vec{f};\vec{\theta}_2)\rangle \quad .\end{aligned} \tag{37}$$

The first two terms on the right-hand-side of Eq. (37) represent the non-isoplanatic dependence which Korff[12] has previously evaluated in connection with speckle interferometry. We shall focus our attention on the last two terms on the right-hand-side of Eq. (37) which, by virtue of isotropy considerations, must be equal. We shall therefore consider

$$\langle \delta_{SE}(\vec{f},\vec{\vartheta})\rangle = \langle \tau_{SE}(\vec{f};\vec{\theta}_1)\,\tau_{SE}^*(\vec{f};\vec{\theta}_2)\rangle \quad . \tag{38}$$

This quantity represents the isoplanatic dependence of short-exposure imagery post-detection compensation in the sense that Eq. (37) can be rewritten as

$$\langle \Delta_{SE}(\vec{f},\vec{\vartheta})\rangle = 2\,[\langle \delta_{SE}(\vec{f},0)\rangle - \langle \delta_{SE}(\vec{f},\vec{\vartheta})\rangle] \quad . \tag{39}$$

We shall seek a convenient formula for the evaluation of $\langle \delta_{SE}(\vec{f},\vec{\vartheta})\rangle$.

Eq. (8) was developed to represent the predetection compensation OTF. It can be adapted to our problem if we replace $\Delta\phi(\vec{r},\vec{\vartheta})$ by $\phi(\vec{r},\vec{\theta})$, corresponding to the replacement of the predetection compensation error, $\Delta\phi$, by the incoming phase error, ϕ , from a point source located in the direction $\vec{\theta}$. For our purposes, we can rewrite Eq. (8) to give the random short-exposure OTF, namely*

$$\tau_{SE}(\vec{f};\vec{\theta}) = A^2 B \int d\vec{r}\; W(\vec{r}-\lambda\vec{f})\, W(\vec{r}) \exp\{i\,[\phi(\vec{r},\vec{\theta}) - \phi(\vec{r}-\lambda\vec{f},\vec{\theta})]\} \quad . \tag{40}$$

If we substitute Eq. (40) into Eq. (38), make a double integral out of the product of integrals, and interchange the order of ensemble averaging and integration, we get

$$\begin{aligned}\langle \delta_{SE}(\vec{f},\vec{\vartheta})\rangle = A^4 B^2 \iint d\vec{r}\, d\vec{r}'\; & W(\vec{r}-\lambda\vec{f})\, W(\vec{r})\, W(\vec{r}'-\lambda\vec{f})\, W(\vec{r}')\, \langle \exp\{i[\phi(\vec{r},\vec{\theta}_1) - \phi(\vec{r}-\lambda\vec{f},\vec{\theta}_1) \\ & - \phi(\vec{r}',\vec{\theta}_2) + \phi(\vec{r}'-\lambda\vec{f},\vec{\theta}_2)]\}\rangle \quad .\end{aligned} \tag{41}$$

Taking advantage of the fact that the argument of the exponential in Eq. (41) is a gaussian random function with zero mean, we can cast the ensemble average in the form

$$\begin{aligned}&\langle \exp\{i[\phi(\vec{r},\vec{\theta}_1) - \phi(\vec{r}-\lambda\vec{f},\vec{\theta}_1) - \phi(\vec{r}',\vec{\theta}_2) + \phi(\vec{r}'-\lambda\vec{f},\vec{\theta}_2)]\}\rangle \\ &= \exp\{-\tfrac{1}{2}\langle [\phi(\vec{r},\vec{\theta}_1) - \phi(\vec{r}-\lambda\vec{f},\vec{\theta}_1) - \phi(\vec{r}_\theta,\vec{\theta}_2) + \phi(\vec{r}'-\lambda\vec{f},\vec{\theta}_2)]^2\rangle\} \quad .\end{aligned} \tag{42}$$

Making use of Eq. (14), we can write

$$\begin{aligned}&\langle [\phi(\vec{r},\vec{\theta}_1) - \phi(\vec{r}-\lambda\vec{f},\vec{\theta}_1) - \phi(\vec{r}',\vec{\theta}_2) + \phi(\vec{r}'-\lambda\vec{f},\vec{\theta}_2)]^2\rangle \\ &= \mathcal{D}_{I,\phi}(\lambda\vec{f},0) + \mathcal{D}_{I,\phi}(\vec{r}-\vec{r}',\vec{\vartheta}) - \mathcal{D}_{I,\phi}(\vec{r}-\vec{r}'+\lambda\vec{f},\vec{\vartheta}) - \mathcal{D}_{I,\phi}(\vec{r}-\vec{r}'-\lambda\vec{f},\vec{\vartheta}) + \mathcal{D}_{I,\phi}(\vec{r}-\vec{r}',\vec{\vartheta}) + \mathcal{D}_{I,\phi}(\lambda\vec{f},0) \\ &= 2\,\mathcal{D}_{I,\phi}(\lambda\vec{f},0) + 2\,\mathcal{D}_{I,\phi}(\vec{r}-\vec{r}',\vec{\vartheta}) - \mathcal{D}_{I,\phi}(\vec{r}-\vec{r}'+\lambda\vec{f},\vec{\vartheta}) - \mathcal{D}_{I,\phi}(\vec{r}-\vec{r}'-\lambda\vec{f},\vec{\vartheta}) \quad .\end{aligned} \tag{43}$$

If we now make use of Eq. (16), and substitute Eq. (43) into Eq. (42), we get

* It should be noted that this formulation suppresses the $\vec{\theta}$-dependent phase shift factor, $\exp(2\pi i \vec{f}\cdot\vec{\theta})$ in the OTF, as though the origin were at the center of the image of the point source. This suppresses an awkward factor, of no physical significance to us, which would otherwise appear when we later combine the transforms of two separately located point sources.

$$\langle \exp\{i[\phi(\vec{r},\vec{\theta}_1) - \phi(\vec{r}-\lambda\vec{f},\vec{\theta}_1) - \phi(\vec{r}',\vec{\theta}_2) + \phi(\vec{r}'-\lambda\vec{f},\vec{\theta}_2)]\}\rangle$$

$$= \exp\left(-2.91\, k^2 \int_{\text{Path}} ds\, C_N^2 \left\{ |\lambda\vec{f}|^{5/3} + |\vec{r}-\vec{r}'+\vec{\vartheta}s|^{5/3} - \tfrac{1}{2}|\vec{r}-\vec{r}'+\lambda\vec{f}+\vec{\vartheta}s|^{5/3} - \tfrac{1}{2}|\vec{r}-\vec{r}'-\lambda\vec{f}+\vec{\vartheta}s|^{5/3}\right\}\right). \quad (44)$$

Now if we substitute Eq. (44) into Eq. (41) and replace the $\vec{r},\vec{r}'$-variables of integration by $\vec{u},\vec{v}$, where $\vec{u}$ and $\vec{v}$ are as defined by Eq.'s (24a) and (24b), and introduce the function $\tilde{W}$, defined by Eq. (25), we get

$$\langle \delta_{SE}(\vec{f},\vec{\vartheta})\rangle = A^4 B^2 D^4 \iint d\vec{u}\, d\vec{v}\, \tilde{W}(\vec{v}+\tfrac{1}{2}\vec{u}-\lambda\vec{f}/D)\, \tilde{W}(\vec{v}+\tfrac{1}{2}\vec{u})\, \tilde{W}(\vec{v}-\tfrac{1}{2}\vec{u}-\lambda\vec{f}/D)\, \tilde{W}(\vec{v}-\tfrac{1}{2}\vec{u})$$

$$\times \exp\left(-2.91\, k^2 D^{5/3} \int_{\text{Path}} ds\, C_N^2 \left\{|\lambda\vec{f}/D|^{5/3} + |\vec{u}+\vec{\vartheta}s/D|^{5/3} - \tfrac{1}{2}|\vec{u}+\lambda\vec{f}/D+\vec{\vartheta}s/D|^{5/3} - \tfrac{1}{2}|\vec{u}-\lambda\vec{f}/D+\vec{\vartheta}s/D|^{5/3}\right\}\right) \quad . \quad (45)$$

It is convenient to define the function $\tilde{K}$ as

$$\tilde{K}(\vec{u},\lambda\vec{f}/D) = \int d\vec{v}\, \tilde{W}(\vec{v}+\tfrac{1}{2}\vec{u}-\lambda\vec{f}/D)\, \tilde{W}(\vec{v}+\tfrac{1}{2}\vec{u})\, \tilde{W}(\vec{v}-\tfrac{1}{2}\vec{u}-\lambda\vec{f}/D)\, \tilde{W}(\vec{v}-\tfrac{1}{2}\vec{u}) \quad . \quad (46)$$

This function has been evaluated by Korff[12].

If we substitute Eq. (46) into Eq. (45), we get

$$\langle \delta_{SE}(\vec{f},\vec{\vartheta})\rangle = A^4 B^2 D^4 \int d\vec{u}\, \tilde{K}(\vec{u},\lambda\vec{f}/D) \exp\left(-2.91\, k^2 D^{5/3} \int_{\text{Path}} ds\, C_N^2 \left\{|\lambda\vec{f}/D|^{5/3} + |\vec{u}+\vec{\vartheta}s/D|^{5/3} - \tfrac{1}{2}|\vec{u}+\lambda\vec{f}/D+\vec{\vartheta}s/D|^{5/3} - |\vec{u}-\lambda\vec{f}/D+\vec{\vartheta}s/D|^{5/3}\right\}\right) \quad . \quad (47)$$

Without making any major approximations, Eq. (47) can not be reduced any further. Fortunately, the expression is susceptible to numerical evaluation without unreasonable effort, given the distribution of C_N^2 . We thus may consider Eq. (47) as our final result, or we may substitute the result into Eq. (37) to obtain Δ_{SE} , the mean square error in post-detection compensation of a short-exposure image. We get in that case

$$\langle \Delta_{SE}(\vec{f},\vec{\vartheta})\rangle = 2\, A^4 B^2 D^4 \int d\vec{u}\, \tilde{K}(\vec{u},\lambda\vec{f}/D) \Big[\exp\left(-2.91\, k^2 D^{5/3} \int_{\text{Path}} ds\, C_N^2 \left\{|\lambda\vec{f}/D|^{5/3} + |\vec{u}|^{5/3} - \tfrac{1}{2}|\vec{u}+\lambda\vec{f}/D|^{5/3} - \tfrac{1}{2}|\vec{u}-\lambda\vec{f}/D|^{5/3}\right\}\right) - \exp\left(-2.91\, k^2 D^{5/3} \int_{\text{Path}} ds\, C_N^2 \left\{|\lambda\vec{f}/D|^{5/3} + |\vec{u}+\vec{\vartheta}s/D|^{5/3} - \tfrac{1}{2}|\vec{u}+\lambda\vec{f}/D+\vec{\vartheta}s/D|^{5/3} - \tfrac{1}{2}|\vec{u}-\lambda\vec{f}/D+\vec{\vartheta}s/D|^{5/3}\right\}\right)\Big] \quad . \quad (48)$$

In developing Eq. (48) from Eq. (37), we have used the fact that the $\vec{u}$-integration treats any value of $\vec{u}$ and its negative equally, so as to allow us to combine the third and fourth terms on the right-hand-side of Eq. (37).

With the results for short-exposure post-detection compensation isoplanatism in hand, as expressed in Eq.'s (47) and (48), we are now ready to turn our attention to the matter of isoplanatism dependence of speckle interferometry. We take this up in the next section.

Speckle Interferometry

The concept of speckle interferometry, introduced by Labeyrie[4], relies on formation of a series of short-exposure images from which, by special processing, information about the object being viewed can be derived with significant information content at spatial frequencies well above that ordinarily allowed by atmospheric turbulence. In fact, the information frequency range is limited only by the diffraction limit of the collector aperture.

The processing introduced by Labeyrie is to calculate the Fourier transform of each of the short-exposure images, square that, and average over the results of the several short-exposure images.* This produces the power spectrum of the image, and this can be associated with the Fourier transform of the correlation function of the object pattern being imaged. As first recognized by Labeyrie, and as demonstrated analytically by Korff, this procedure limits the ability of atmospheric turbulence to wash out the high spatial frequency information anywhere near as completely as it does in the case of ordinary imagery. In imaging a pair of equal intensity point sources using speckle interferometry, the derived

* These short-exposure images appear speckled — hence the name "speckle interferometry."

power spectrum will appear as an approximately gaussian tapered pattern whose width is inversely proportional to the turbulence-limited ordinary imagery resolution. Superimposed on this pattern will be a sinusoidal modulation with a characteristic period that is inversely proportional to the angular separation of the two point sources.

The proper measure of the performance of speckle interferometry is the contrast of the sinusoidal modulation. Korff evaluated this modulation assuming that the signal from both point sources experienced the same turbulence-induced wavefront distortion. We shall here revise the formulation to take proper account of the fact that the two wavefronts, coming from somewhat different directions, experience somewhat different distortions. This constitutes the isoplanatism dependence of speckle interferometry.

We consider point sources at $\vec{\theta}_1$ and $\vec{\theta}_2$, producing random short-exposure intensity patterns in the focal plane $I(\vec{x};\vec{\theta}_1)$ and $I(\vec{x};\vec{\theta}_2)$. The composite random image we shall be dealing with is

$$I_T(\vec{x}) = I(\vec{x};\vec{\theta}_1) + I(\vec{x};\vec{\theta}_2) \quad . \tag{49}$$

The transform of each point image can be written as

$$\tau(\vec{f};\vec{\theta}_1) = B \int d\vec{x}\, I(\vec{x};\vec{\theta}_1) \exp(2\pi i \vec{f}\cdot\vec{x}) \quad , \tag{50a}$$

$$\tau(\vec{f};\vec{\theta}_2) = B \int d\vec{x}\, I(\vec{x};\vec{\theta}_2) \exp(2\pi i \vec{f}\cdot\vec{x}) \quad , \tag{50b}$$

which, because of our choice of the constant of proportionality, B, we can identify as the instantaneous random OTF's associated with the two directions, $\vec{\theta}_1$ and $\vec{\theta}_2$. The transform of the composite image is

$$\mathscr{I}_T(\vec{f};\vec{\vartheta}) = \int d\vec{x}\, I_T(\vec{x}) \exp(2\pi i \vec{f}\cdot\vec{x}) \quad . \tag{51}$$

If we substitute Eq.'s (49) and (50a and b) into Eq. (51), we get

$$\mathscr{I}_T(\vec{f};\vec{\vartheta}) = B^{-1}[\tau(\vec{f};\vec{\theta}_1) + \tau(\vec{f};\vec{\theta}_2)] \quad . \tag{52}$$

The ensemble average of the square of the transform is

$$\langle|\mathscr{I}_T(\vec{f};\vec{\vartheta}|^2\rangle = B^{-2}\,\langle|\tau(\vec{f};\vec{\theta}_1)|^2\rangle + \langle|\tau(\vec{f};\vec{\theta}_2|^2\rangle + \langle\tau^*(\vec{f};\vec{\theta}_1)\,\tau(\vec{f};\vec{\theta}_2)\rangle + \langle\tau^*(\vec{f};\vec{\theta}_2)\,\tau(\vec{f};\vec{\theta}_1)\rangle \,. \tag{53}$$

This equation requires some physical interpretation before we can proceed further. We note that the first two terms on the right-hand-side are essentially independent of $\vec{\theta}_1$ and $\vec{\theta}_2$. The second two terms can be shown to give rise to a sinusoidal oscillation in the $\vec{f}$-plane of the form $\cos(2\pi\vec{f}\cdot\vec{\vartheta})$. This arises from the phase shift in the Fourier transform associated with the displacement of the centroid of each image to the focal plane locations associated with $\vec{\theta}_1$ and $\vec{\theta}_2$.

If we interpret $\tau(\vec{f};\vec{\theta})$ in accordance with the formulation given in Eq. (40), suppressing the phase shift * then the amplitude of the oscillatory component can be written as

$$A(\vec{f},\vec{\vartheta}) = \langle\tau^*(\vec{f};\vec{\theta}_1)\,\tau(\vec{f};\vec{\theta}_2)\rangle + \langle\tau^*(\vec{f};\vec{\theta}_2)\,\tau(\vec{f};\vec{\theta}_1)\rangle \quad . \tag{54}$$

This represents the isoplanatism dependent signal hardening capability of speckle interferometry. If we compare this with Eq. (38), we see that this result is essentially identical to the short-exposure post-detection compensation isoplanatism dependence, $\langle\delta_{SE}(\vec{f},\vec{\vartheta})\rangle$.

$$A(\vec{f},\vec{\vartheta}) = 2\,\langle\delta_{SE}(\vec{f},\vec{\vartheta})\rangle \quad . \tag{55}$$

This completes our analysis of isoplanatism dependencies for various optical processes. In the next section, we shall gather all of these results together and comment on their interrelationships.

Comparison of Results

It is appropriate at this point to gather together our various results for isoplanatism for intercomparison. With the exception of the long-exposure post-detection compensation result, that the discrepancy in the compensated OTF will have an average value of

$$\Delta_{LE}(f,\delta\theta) = \{\tau_{LE}(f)\,\ell n[\tau_{LE}(f)]\tan\theta\,\delta\theta\}^2 \quad , \tag{35}$$

* cf. Footnote in previous section.

all of the other results bear a strong similarity. This suggests that the dependence on the isoplanatic angle, ϑ , will be generally the same. However, the differences between the equations are, as we may note by comparing the following, sufficiently great that we could hardly expect anything like equivalence between the various isoplanatism dependencies.

For predetection compensation, the isoplanatism dependence can be written as for the achieved OTF,

$$M(\lambda\vec{f},\vec{\vartheta}) = \exp\left[-\mathcal{m}(\lambda\vec{f},\vec{\vartheta})\right] \qquad , \tag{12}$$

where

$$\mathcal{m}(\lambda\vec{f},\vec{\vartheta}) = 2.91\, k^2 \int_{Path} ds\, C_N^2 \{(\vartheta s)^{5/3} + (\lambda f)^{5/3} - \tfrac{1}{2}[(\lambda f)^2 + 2(\lambda f)(\vartheta s)\cos\phi + (\vartheta s)^2]^{5/6}$$
$$- \tfrac{1}{2}[(\lambda f)^2 - 2(\lambda f)(\vartheta s)\cos\phi + (\vartheta s)^2]^{5/6}\} \quad . \tag{17}$$

For angle-of-arrival variations, the isoplanatism dependence of the mean square difference in angle-of-arrival is

$$\mathcal{D}_{I,\alpha}(\vec{\vartheta}) = 2.91\,(16/\pi)^2 D^{-1/3} \int_{Path} ds\, C_N^2 \int_0^{2\pi} d\phi \int_0^1 u\, du\, [\cos^{-1}(u) - (3u-2u^2)(1-u^2)^{1/2}]$$
$$\times \{-(u)^{5/3} - (\vartheta s/D)^{5/3} + \tfrac{1}{2}[u^2 + 2u(\vartheta s/D)\cos\phi + (\vartheta s/D)^2]^{5/6} + \tfrac{1}{2}[u^2 - 2u(\vartheta s/D)\cos\phi$$
$$+ (\vartheta s/D)^2]^{5/6}\} \quad . \tag{28}$$

For short-exposure post-detection compensation, the correlation of the random OTF's is

$$\langle \delta_{SE}(\vec{f},\vec{\vartheta})\rangle = (A^4 B^2)\, D^4 \int d\vec{u}\, \tilde{K}(\vec{u},\lambda\vec{f}/D) \exp\Big(-2.91\, k^2 D^{5/3} \int_{Path} ds\, C_N^2 \{|\lambda\vec{f}/D|^{5/3} + |\vec{u}+\vec{\vartheta}s/D|^{5/3}$$
$$- \tfrac{1}{2}|\vec{u}+\lambda\vec{f}/D+\vec{\vartheta}s/D|^{5/3} - \tfrac{1}{2}|\vec{u}-\lambda\vec{f}/D+\vec{\vartheta}s/D|^{5/3}\}\Big) \quad . \tag{47}$$

For speckle interferometry, the amplitude of the oscillating component of the speckle pattern's Fourier transform is

$$A(\vec{f},\vec{\vartheta}) = 2\,\langle \delta_{SE}(\vec{f},\vec{\vartheta})\rangle \tag{55}$$

The identical nature of the isoplanatism dependence of speckle interferometry and of short-exposure imagery is obvious and needs no special comment. On the other hand, while there are obvious similarities between the isoplanatism dependencies for predetection compensation, for angle-of-arrival correction, and for short-exposure post-detection compensation, the dependencies are clearly distinct. We would expect the dependencies to all be of the same order of magnitude, but would not be surprised to find as much as a factor of two difference in the field-of-view isoplanatism limitations for these different processes. We conclude that to be properly meaningful, the term "atmospheric turbulence isoplanatism" requires a modifier defining the type of optical process of interest — that measurement of the isoplanatism dependence of one type of optical process does not generally provide better than order of magnitude information about the isoplanatism dependence of another type of optical process.

References

1. M. Born and E. Wolf, "Principles of Optics," III Edition, Pergamon Press, New York (1965); p. 482.
2. W. B. Bridges, et al, "Coherent Optical Adaptive Techniques," Appl. Opt. 13, 291 (1974)
3. J. L. Harris, Sr., "Image Evaluation and Restoration," J. Opt. Soc. Am. 56, 569 (1966)
4. A. Labeyrie, Astron. Astrophys. 6, 85 (1970)
5. I am indebted to Dr. Merlin Miller of AVCO-Everett Research Laboratory for pointing out to me how easily the isoplanatic dependence of speckle interferometry could be related to that for post-detection compensation of a short-exposure image.
6. R. F. Lutomirski and R. G. Buser, "Mutual Coherence Function of a Finite Optical Beam and Application to Coherent Detection," Appl. Opt. 12, 2153 (1973). To obtain the result presented here, we have adapted the results of the reference by treating the inner and outer scales of turbulence at zero and infinity, respectively, and then performed the spatial frequency integration. We then further assumed that the entire turbulent portion of the propagation path was much closer to the receiver than to the source, as it would be in viewing a space object.

7. D. L. Fried, "Isoplanatism in Predetection Compensation Imaging," in Topical Meeting on Optical Propagation Through Turbulence, Boulder, Colo., July 9-11, 1974. (It should be noted that because of interpretation of a half-angle as a full-angle, there is a factor of two error in the results in this reference. The isoplanatic patch size is twice as large as is implied in the reference.)

8. D. L. Fried, "Statistics of a Geometric Representation of Wavefront Distortion," J. Opt. Soc. Am. 55, 1427 (1965). It can be seen from this reference that so long as an aperture's diameter is less than about 3 r_0, it is reasonably accurate to consider the wavefront distortion over the aperture to be little more than a random tilt.

9. D. L. Fried and H. T. Yura, "Telescope-Performance Reciprocity for Propagation in a Turbulent Medium," J. Opt. Soc. Am. 62, 600 (1972)

10. D. L. Fried, "Differential Angle of Arrival: Theory, Evaluation, and Measurement Feasibility," Rad. Sci. 10, 71 (1975)

11. D. L. Fried, "Optical Resolution Through a Randomly Inhomogeneous Medium for Very Long and Very Short Exposures," J. Opt. Soc. Am. 56, 1372 (1966)

12. D. Korff, "Analysis of a Method for Obtaining Near-Diffraction-Limited Information in the Presence of Atmospheric Turbulence," J. Opt. Soc. Am. 63, 971 (1973).

CHARACTERIZATION OF ATMOSPHERIC TURBULENCE*

M. Miller
Avco Everett Research Laboratory, Inc., Everett, MA

P. Zieske
Avco Everett Research Laboratory, Inc., Puunene, Maui, HA

D. Hanson
Rome Air Development Center
Griffiss Air Force Base, NY

Abstract

Measurements were made of the atmospheric temperature turbulence surrounding and above the ARPA Maui Optical Station (AMOS) using several advanced remote turbulence sensors. Simultaneously, optical propagation parameters were measured directly. A brief description of the various instruments and their measurement capabilities is given. Typical atmospheric turbulence data for the measurement period are presented. Comparisons between optical propagation parameters measured directly and those calculated from the measured turbulence are presented.

Introduction

A major influence on the operation of large optical systems arises from random phase and amplitude perturbations introduced into a propagating optical field by atmospheric turbulence. The origin of these effects are local small scale, temperature fluctuations which, in turn, produce variations in the index of refraction. These random perturbations in the received field are very significant in imaging systems because they limit resolution to the order of one or more arc seconds. A variety of techniques(1) - (6) have been proposed and in some cases implemented which minimize these effects to some degree. Experimental data is desirable for the evaluation, design and implementation of these systems.

In order to satisfy this data requirement, a set of atmospheric turbulence characterization experiments have been developed and implemented. Included are a variety of instrumentation which measure properties of the turbulent environment and its effects on optical propagation. The site of these measurements is the ARPA Maui Optical Station (AMOS) atop Haleakala on the island of Maui, Hawaii. The long term objective is to establish a data base on the long and short term statistics of atmospheric seeing conditions. As part of this program, an intensive series of measurements were carried out during November and December of 1975. This paper reports the initial results of these measurements.

During this intensive measurement period, the primary goals were: 1) to estimate the locations and strengths of regions of high turbulence and 2) to measure the mutual coherence function correlation scale (r_o) and its short term statistics. Four instrumental systems were used to obtain data relative to the first of these goals. Reduction of these data results in an estimate of the atmospheric refractive-index structure constant, C_n^2, at a variety of altitudes above the observatory. Estimates of r_o were obtained by use of an instrument which measures properties of the short exposure image of a stellar source. Numerical comparisons between these two types of measurements were also carried out.

Brief descriptions of the various instruments are given in the next section. This is followed by a description of the data collection, reduction and processing procedures. The final two sections of this paper summarize the reduced data and the numerical comparisons.

I. Measurement Systems

Micrometeorological Instrumentation

Microthermal sensors(7) were used to obtain an estimate of the near-ground level strength of turbulence. They measure small scale high frequency ($\simeq$ 100 Hz) temperature fluctuations using thin wire (1.27×10^{-3} cm diameter) probes. Three of these sensors were mounted in a triangular array atop two 20 meter towers which are located upwind and downwind of the observatory during typical tradewind patterns. Cup anemometers and wind vanes were located on the towers to provide wind speed and direction. An ambient and dew point thermometer were also attached to one of the towers.

When taken in pairs, the microthermal sensors provide a direct measure of the temperature structure constant, C_T^2. These data can be converted to C_n^2 values if the average temperature and pressure is known as shown in Eq. (1).

*This work was sponsored in part by The Defense Advanced Research Project Agency and the Air Force Systems Command, Rome Air Development Center under Contract F30602-76-C-0054.

$$C_n^2 = (4.76 \times 10^{-9})\ (\bar{p}^2/\bar{T}^4)\ (\overline{\Delta T^2}/R^{2/3}) \qquad (1)$$

where $\bar{p}$ is the average pressure, $\bar{T}$ is the average temperature, R is the probe spacing (1 meter) and $\overline{\Delta T^2}$ is the mean-squared temperature difference of the two probes. Hence, this system provides six values of C_n^2 (three from each tower). However, depending on wind direction, one value from each array should be a more accurate estimate.

Airborne Microthermal Probe

During a portion of the measurement period, personnel from the NOAA Environmental Research Laboratories flew a light aircraft instrumented with a microthermal probe[8] over the observatory. The resulting data were reduced by NOAA yielding estimates of C_n^2 from approximately 40 to 2,000 meters above the site.

Acoustic Sounder

This device[9] (built by NOAA Environmental Research Laboratory) is essentially a monostatic acoustic radar which measures range-gated returns from temperature variations. The return power is recorded qualitatively on a facsimile recorder. Provided temperature and humidity information is available, quantitatively recorded data can be reduced for estimates of C_n^2 from approximately 30 to 300 meters with a 30 meter resolution size. Due to a large acceptance angle, data taken lower than 30 meters is very susceptible to local reflections from building, etc.

Star Sensor

This instrument (built by NOAA Environmental Research Laboratory) consists of a 35.6 cm Schmidt-Cassegrain telescope whose aperture plane is imaged approximately on a variable spatial filter.[10] When a stellar source is viewed through this system, measurements of the spatial frequency content of the atmospherically produced scintillation pattern are obtained. The resulting data is read and processed by a dedicated NOVA 210 computer.

The general philosophy of the operation of this instrument is as follows. The signal at a specific spatial frequency depends not only on the strength of turbulence, but also on the location. By combining signals at various spatial frequencies in an appropriate fashion, weighing functions centered at different heights can be obtained. Hence, direct optical measurements are used to obtain estimates of C_n^2 centered at approximately 2.5, 3, 5, 7.5, 10, 12.5, and 25 km from the devices. These data processing algorithms are carried out in real time by the on-line computer once each twenty minutes. While the raw data represents a direct optical measurement, the reduction of these data for values of C_n^2 requires the assumption of a model of the turbulence and its effect on optical propagation.[10] Departures from this model will result in incorrect values of C_n^2 vs height.

In addition to the processed structure constant data, the star sensor also records the aperture plane log-amplitude variance,

$$\sigma_\ell^2 = \frac{1}{4} \ln (\sigma_p^2 + 1) \qquad (2)$$

where

$$\sigma_p^2 = \frac{\overline{(p - \bar{p})^2}}{\bar{p}^2}$$

and p is the integrated irradiance over 35.6 cm, centrally obscured (obstruction ration of ~ .3) aperture. A one-second analog average of this quantity is calculated and sampled. After collecting twenty-four of these samples, the computer calculates a sample average and prints out the results. During the twenty minute cycle required for structure constant data, approximately 40 values of this intermediate average are obtained. A final calculation results in a twenty minute mean and standard deviation of the log-amplitude variance. The latter quantity yields a measure of the stationarity of the data over each measurement cycle and is important in judging the validity of the structure function data. Large values of the standard deviation do not in principle, invalidate the log-variance data. However, experience has shown that very large values indicate instrument malfunction or the existance of partial cloud cover during the measurement cycle.

Seeing Monitor

The Seeing Monitor[11] (built by Hughes Research Laboratory) is an instrument used to measure the spatial frequency content of a stellar source as imaged by a telescope of arbitrary aperture. For these experiments, it was mounted on a 1.2 meter diameter telescope which had an F number of 16. The measurement is made by imaging the star on a spinning, variable spatial frequency reticle in front of a photomultiplier. The amplitude of the PM signal is proportional to the combined telescope and atmosphere modulation transfer function (MTF). This system is duplicated yielding simultaneous measurements along two orthogonal directions. A scan is completed in one millisecond and is repeated continuously.

While a variety of outputs are available, the ones used in these experiments were the two seeing angles which are related to the spatial frequency (in angular units) at which the MTF has dropped to one-half of its initial value. Averages of these quantities can be used to estimate the mutual coherence function correlation scale (r_o) by fitting to the theoretical expression[12]

$$MTF(\lambda f) = t_o(\lambda f) \exp -3.44 (\lambda f/r_o)^{5/3} \left[1 - \alpha(\lambda f/D)^{1/3}\right] \quad (3)$$

where λ is wavelength, f is spatial frequency, D is the aperture diameter and t_o is the telescope MTF. α is a parameter whose value is 1.0 for $D^2 \gg \lambda L$ (near field) and 0.5 for $D^2 \ll \lambda L$ (far field) where L is the length of the turbulent path. The averaged short exposure MTF (Eq. (3)) has been used because overall image wander is removed during the measurement process.

II. Experimental Configuration

The micrometeorological system provides a total of twelve simultaneous and continuous analog outputs which require processing to obtain the information of interest. These signals were processed in real time using an on-line computer system.

The data processing system consists of a PDP-8I computer, teletype, paper tape reader and a 16-channel, selectable gain, bi-polar, 10 bit A/D converter. The conversion aperture of the computer is 0.1 μs and its maximum throughput rate is 34 kHz. In actual operation, the throughput rate is limited by software rather than hardware. The PDP-8 is programmed to calculate the mean and variance of each channel and the covariance matrix for the entire array of inputs. Within physical constraints, the number of channels, sampling rate and number of samples can be varied. After obtaining the required number of samples, data collection is interrupted in order to accommodated data printout. At the completion of printout, the system automatically starts a new collection cycle. The system continues to cycle in this fashion until it is manually halted.

For the measurements reported here, 1350 samples of each input were included in a single cycle. This required approximately 8.25 minutes for data collection and 1.75 minutes for printout. A basic 10 minute cycle was selected in order to provide a convenient synchronization with the twenty minute star sensor cycle. Relevant real time outputs for the microthermal system included temperature variances (6), temperature covariances (6) and average temperature for each cycle. Average pressure was recorded using a mercury barometer located in the observatory. Reduction of these data for values of C_n^2 was carried out at a later time using Eq. (1).

The data processing system was also used to process the two analog seeing angle outputs from the Seeing Monitor. Hence, estimates of the mean and variance of these two quantities based on 1350 samples were obtained. Conversion of these results to physical units and estimates of r_o via Eq. (3) were made at a later time. For all results reported, a spectral filter with a bandwidth of 1200 Å centered at 6300 Å was placed in the Seeing Monitor optical train. Stars of approximate visual magnitude + 2 and elevations of 60° or greater were used throughout the measurement period.

As discussed in the previous section, the Star Sensor is a completely independent system which collects and processes data in real time. Once initial alignment was established, all that was required was occasional fine adjustments of the track. Stars of approximate visual magnitude + 2 at elevations of 60° or higher were used as sources. Spectral characteristics were determined by the S-20 response of the photomultipliers.

Data was obtained with this instrumentation during the period 11 November to 8 December 1975. Seven three-hour data runs were collected with all systems operational. Five of these were started in the 1900-2300 time period. The other two were started in the 0000-0200 time period. Three additional data runs with an incomplete complement of instrumentation were also collected.

During four of these data runs (17-21 November), NOAA personnel collected microthermal data using the airborne sensor. The procedure used was to begin operations with the ground-based instrument before the aircraft arrived on station. During the data run, the aircraft made several spiral decents from approximately 2000 m above the observatory. Sufficient time marks were made to allow approximate time synchronization of the data. Airborne data was reduced by NOAA at their Boulder facility.

III. Results

Seeing Monitor

A total of 87 data points were collected during ten nights. Each represented a ten minute average of the two seeing angle outputs. A summary of these data is given in Table I. The mean voltage is an average over the samples taken during each night. The range of values recorded during each night is also included. The voltage output, V, is related to the one-half MTF spatial frequency by the calibration expression.

$$f_{1/2} \simeq 0.035 \exp. (0.46V) \text{ (arc sec)}^{-1} \quad (4)$$

The values of r_o reported in Table I were obtained in the following way. For each pair of data points, values of r_o were calculated using Eq. (3) assuming a clear, diffraction limited aperture of 1.2m, a wavelength of 6300 Å and near-field turbulence conditions ($\alpha = 1$). These two values were averaged and then corrected for zenith angle dependence using the theoretical expression[12]

$$r_o(0) = (\sec \Theta)^{3/5}\, r_o(\Theta) \tag{5}$$

where $r_o(\Theta)$ is the correlation scale at zenith angle Θ. Average values over an entire data run were obtained by use of the expression

$$r_o = \left[\frac{1}{N} \sum_{i=1}^{N} (r_{oi})^{3/5} \right]^{5/3} \tag{6}$$

The three-fifths power of r_o was averaged in order to obtain a more direct comparison with expressions developed from the experimentally derived turbulence profile data.

For the entire 174 data points (87 in each channel) the average voltage was 5.82V with a range of (4.28-7.22)V. The average r_o (at 6300 Å) for the entire experimental period was 13.6 cm with a range of (8.8-24.7) cm. Assuming the theoretical scaling of r_o with wavelength ($\lambda^{6/5}$) results in an average r_o of 10.3 cm with a range of (6.7-18.7) cm at 5000 Å. These values are consistent with previous measurements made at AMOS[13], [14] and other sites.[15]

Digital processing of the raw data was done with a 10 bit A to D converter with a full scale range of 10 volts. Therefore, the precision of the measurements is approximately 0.01 volts. Using Eqs. (3) and (4), this results in a precision in r_o of approximately 0.1 cm. Noise sources in the system are as yet, not fully evaluated. Therefore, it is difficult at this time to estimate the accuracy of the results. However, the data processing system also calculated the 1350 sample variance in each basic ten minute cycle. Standard deviations obtained were in the range (6-20)% of the mean with a typical value of 10%. Assuming independent samples, this yields an estimated standard deviation of the sample mean of less than 0.5%. Therefore, the statistical significance of the sample mean should be of the same order of magnitude as the precision of the measurements. Further experiments are required in order to determine the relative fractions of the total variance which are due to noise sources and the atmosphere, respectively, as well as any influence of noise on the measured mean value.

Star Sensor Log Amplitude Variance

A summary of the twenty minute log amplitude variance is given in Table II. The number of cycles, average and range for each night are included. The theoretical zenith angle dependence has not been removed. Because all data was collected on stars at low zenith angle (typically 10-20°, 30° maximum), this effect should not be large. The maximum twenty minute normalized standard deviation for the data included is 50%, with most values below 20%. For the entire 88 samples included in the table, the range is 1.55×10^{-4} to 2.83×10^{-3} with an average of 5.51×10^{-4}.

A characterization of the noise contribution to this data has not yet been carried out. Therefore, the accuracy cannot be estimated. The precision is approximately equal to the three significant figures shown in the table. Because of the manner in which the data is processed, it is difficult to estimate its true level of statistical significance. However, for the mean value of the twenty-four second averages, the data indicates a statistical significance in the range of (1-8)%.

Turbulence Profile

Turbulence profiles averaged over the complete data runs taken on 17, 18 and 21 November are shown in Figures 1, 2 and 3. The lowest height (15 m) data points were obtained from the ground based microthermal sensors. Data from 37 m to 2.5 km were provided by NOAA from a reduction of their airborne microthermal data. The line segments from 1 km to 24 km were derived from Star Sensor data and represent the approximate width of the wieghing functions used in the data reduction.[10] The horizontal scale is in height above the observatory which is at an altitude of approximately 10,000 ft. The tropopause height was determined from rawinsonde temperature profiles obtained from the two Hawaiian Weather Bureau stations.

Values of r_o and σ_1^2 (point aperture) were calculated from these profiles using the expressions[12], [16]

$$r_o = \left[0.42\, (2\pi/\lambda)^2 \int_{path} dh\, C_n^2 (h) \right]^{-3/5} \tag{7}$$

$$\sigma_1^2 = 0.56\, (2\pi/\lambda)^2 \int_{path} dh\, h^{5/6}\, C_n^2 (h) \tag{8}$$

A hewlett Packard 9810 desk top calculator with a peripheral digitizer was used to carry out these calculations. Programming involved forming the Riemann sum using 45 points between 15 m and 25 km. In the overlap region, Star Sensor data was used. The results of these numerical calculations are given in Table III.

IV. Discussion

In Table IV we have tabulated the values of r_o calculated from the turbulence profiles and derived from the Seeing Monitor data. Also included is the factor by which the C_n^2 profile integral would have to be multiplied to obtain agreement between the two values.

The values of σ_1^2 given in Tables II and III cannot be directly compared because the experimental data of Table III corresponds to a 14 inch aperture while the derived values of Table III represent point scintillation. Detailed comparisons between the profile and Star Sensor data are reported by Ochs, Wang and Lawrence.[10] The agreement is good.

As can be seen from these results, agreement between the profile and Seeing Monitor data is not good. A possible explanation for this is that significant amounts of turbulence was missed by the profiling instrumentation. Assuming this to be the case, the good agreement of the scintillation data would indicate that the missing turbulence is at low altitudes. It should be noted that the low altitude airborne data was not taken directly over the observatory but over a region on a lower portion of the mountain. Therefore, the data obtained may not be representative of conditions at the observatory because the relative distance to the ground was greater.

The required increase of the C_n^2 integral to bring the two values into agreement could, for example, be accounted for by a turbulent layer of order 20-80 meter thick of average strength $C_n^2 = 10^{-14}\ m^{-2/3}$. These values are not unreasonable. If such a layer existed in the first 1000 meters above the observatory, it could have been seen by the Acoustic Sounder. Unfortunately, quantitative reduction of this data has not been successful due to problems with the instrumental operation and/or processing software.

In the future we plan to continue these data collection activities. Our primary objective are to obtain a more quantitative and detailed measure of the turbulence profile and a characterization of the long term statistics of atmospheric seeing conditions.

Acknowledgment

We would like to thank the staff of the AMOS Observatory, and in particular R. Taft and W. Zane for assistance in collecting the data. J. Spencer and D. Tarazano of RADC assisted in instrument installation, data reduction and contributed to many useful discussions. We acknowledge several discussions with J. Ochs and R. Lawrence of NOAA relative to the Star Sensor and thank them for providing us with the results of their airborne experiments. D. Fried of Optical Science Consultants contributed to discussions covering many aspects of work reported here.

References

1. Labeyrie, A., Astronomy and Astrophysics 6, 85 (1970).

2. Currie, D. C., Topical Meeting on Imaging in Astronomy, Cambridge, Mass. (1975).

3. Muller, R. A. and Buffington, A., J. Opt. Soc. Am. 64, 1200 (1974).

4. Miller, L., Brown, W. P., Jenney, J. A. and O'Meara, T. R., OSA Topical Meeting on Optical Propagation Through Turbulence, Boulder, Colorado (1974).

5. Hardy, J. W., Feinleib, J. and Wyant, J. C., OSA Topical Meeting on Optical Propagation Through Turbulence, Boulder, Colorado (1974).

6. Yellin, SPIE Symposium on Imaging Through the Atmosphere, Reston, Virginia (1976).

7. Greenwood, D. P. and Youmans, D. B., A Fine-Wire Microtemperature Probe for Atmospheric Measurements, RADC-TR-75-240 (Griffis AFB, December 1975).

8. Lawrence, R. S., Ochs, G. R. and Clifford, S. F., J. Opt. Soc. Am. 60, 826 (1971).

9. Hall, F. F., Jr., in Temperature and Wind Structure Studies by Acoustic Echo Sounding, ed. by V. E. Derr, U. S. Govt. Pr. Off. 0322-0011 (1972).

10. Ochs, G. R., Wang, Ting-i and Lawrence, R. S., SPIE Symposium on Imaging Through the Atmosphere, Reston, Virginia (1976).

11. Giuliano, C. R. et al, Space Object Imaging, Final Report, Contract F30602-74-C-0227 (Rome Air Development Center, Griffiss AFB, New York).

12. Fried, D. L., J. Opt. Soc. Am. 56, 1372 (1966).

13. Miller, M. G. and Kellen, P. F., Topical Meeting on Imaging in Astronomy, Cambridge, Mass. (1975).

14. Schneiderman, A. M. and Karo, D., OSA Topical Meeting on Speckle Phenomena in Optics, Microwaves and Acoustics, Pacific Grove, California (1976).

15. Dainty, J. C. and Scadden, R. J., Mon. Not. R. Astr. Soc. 170, 519 (1975).

16. Lawrence, R. and Strohbehn, J., Proc. IEEE 58, 1523 (1970).

TABLE I

SEEING MONITOR DATA

Date	Time		No. of	Output Voltage				r_0 (cm)	
				Channel #1		Channel #2			
	Start	End	Sam.	Mean	Range	Mean	Range	Mean	Range
Nov. 11	2243	0043	6	5.83	5.44-6.20	5.56	5.12-6.00	12.4	10.3-14.8
Nov. 12	1948	2248	9	5.69	5.28-5.98	5.86	5.51-6.14	12.8	10.8-14.5
Nov. 14	2110	0010	9	6.22	6.05-6.45	6.07	5.94-6.24	15.8	14.7-17.3
Nov. 15	1927	2227	10	6.14	5.88-6.44	6.05	5.82-6.27	15.6	14.0-18.2
Nov. 17	1945	2245	9	5.17	4.87-5.72	5.19	4.86-5.79	10.4	9.3-13.5
Nov. 18	2050	2350	9	6.06	5.66-6.48	6.10	5.73-6.42	15.4	13.3-18.2
Nov. 19	2015	2315	9	5.47	4.99-5.84	5.44	4.97-5.76	11.8	9.4-14.4
Nov. 21	1929	2229	8	5.44	5.33-5.92	5.49	4.82-5.89	11.6	8.8-14.1
Dec. 6	0055	0335	9	5.72	5.85-5.54	5.59	5.45-5.71	12.7	11.9-13.3
Dec. 8	0207	0507	9	6.61	5.90-7.22	6.45	5.68-7.07	18.3	13.6-24.7

TABLE II

STAR SENSOR LOG-AMPLIFIER VARIANCE

Date	Time		No. of Samples	σ_ℓ^2	
	Start	End		Range $\times 10^{-4}$	Average $\times 10^{-4}$
Nov. 12	1943	2248	2	5.50-5.54	5.52
Nov. 14	1910	0010	15	2.78-6.22	4.16
Nov. 15	1807	2227	13	1.55-3.98	2.68
Nov. 17	1845	2245	12	2.79-18.2	7.92
Nov. 18	1950	2350	12	2.28-5.03	3.76
Nov. 19	1935	2315	4	2.04-28.3	18.3
Nov. 21	1929	2229	8	5.17-8.21	6.00
Dec. 6	0000	0335	12	4.51-10.4	6.88
Dec. 8	0147	0507	10	2.13-5.71	3.24

TABLE III

Turbulence Profile Parameters

Date	Correlation Scale r_o (cm)	Log-Amplitude Variance σ_1^2
17 November	25.7	0.018
18 November	24.9	0.008
21 November	16.6	0.019

TABLE IV

Data Comparisons

Date	Seeing Monitor r_o (cm)	Profile r_o (cm)	Profile Integral Multiplier
17 November	10.4	25.7	4.5
18 November	15.4	24.9	2.2
21 November	11.6	16.6	1.8

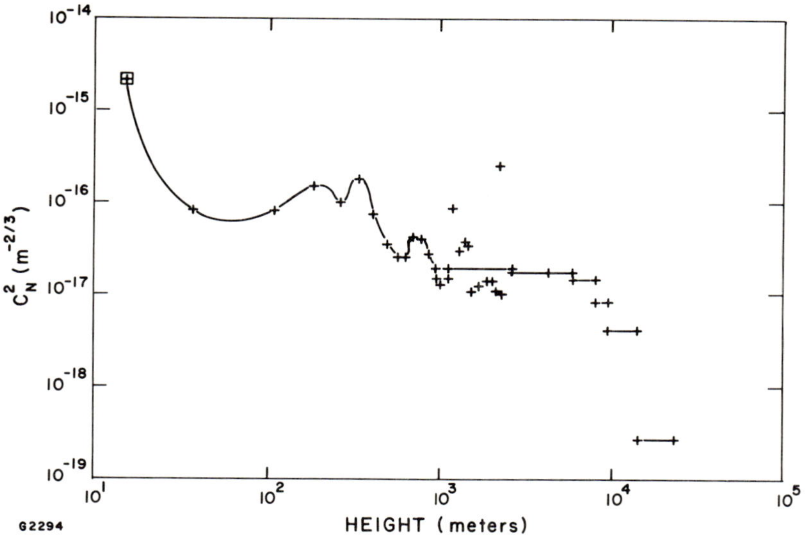

Figure 1: Turbulence Profile vs Height Above the Observatory - 17 November 1975. Bround Based Microthermal Data = □; Airborne Microthermal Data = +; Star Sensor Data = +—+. Estimated tropopause height of 11-15.5 km.

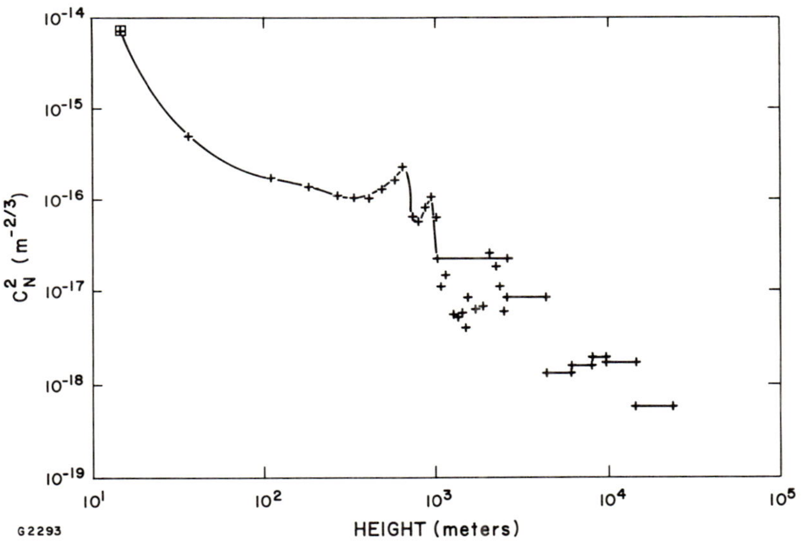

Figure 2: Same as Figure 1 for 18 Novermber 1975. Estimated tropopause height of 9-15.5 km.

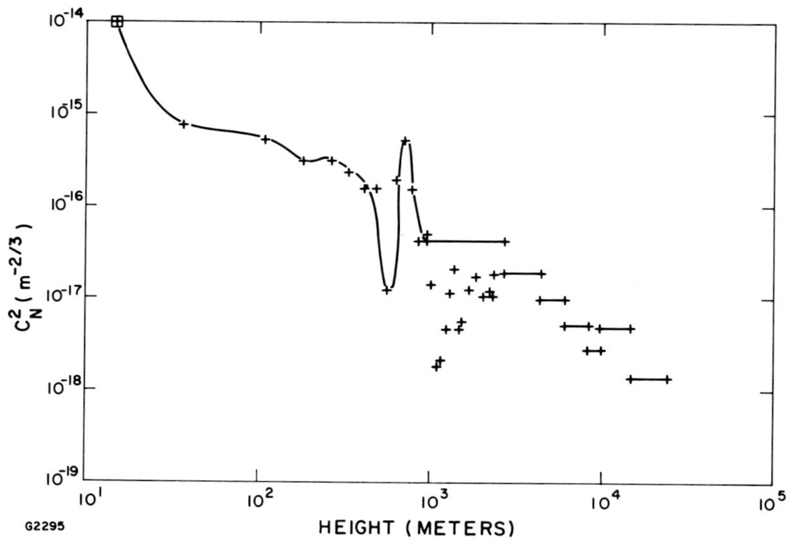

Figure 3: Same as Figure 1 for 21 November 1975. Estimated tropopause height of 9.1-15 km.

DYNAMIC ATMOSPHERIC TURBULENCE CORRECTIONS

Robert J. Noll
The Perkin-Elmer Corporation
Norwalk, Connecticut 06856

Abstract

The wavefront distortions produced by atmospheric fluctuations are discussed in this paper. The problem at hand is: What is the best way to process the measurements of these distortions so that appropriate corrections for them can be made? If a set of N independent wavefront measurements are made, the measured wavefront can be established as some linear combination of these measurements. The measurements themselves need not be direct phase measurements but could be a set of wavefront slope measurements. Nevertheless, the problem is to find a procedure that gives a best estimate of the wavefront from the set of N measurements. With such a procedure, the system designer can make an estimate of the number of measurements required to achieve a certain desired level of performance as well as the dynamical system complexity required to process the data. What is considered here is an application and adaptation of the theory of optimal estimates to the problem of random wavefront estimation.

General System Description

A generic image compensation system is composed of three basic elements: a wavefront sensor, a wavefront corrector, and a data processor to link the two. The interaction of these three components may be viewed as a servo system. This paper discusses such a system with emphasis on guidelines that can be followed by the system designer in the design of an image compensation system.

If we limit ourselves to linear systems, the wavefront sensor can be generally described by the relation:

$$P(j,t) = P_o(j) + \sum_{\ell} B_{j,\ell}\, S(\ell,t) + n(j,t) \qquad (1)$$

Equation 1 states that the photon flux, $P(j,t)$, measured by the wavefront sensor at some place, j, and at time, t, is the sum of the flux, $P_o(j)$, with no phase disturbance plus the effect of the phase fluctuation. The precise nature of the matrix (B) is defined by the specific wavefront sensor to be employed. The vector, S, denotes the net phase displacement due to the sum of the turbulence and the wavefront corrector at a point, ℓ, in the wavefront measurement plane. The term $n(j,t)$ appears, in Eq. 1, to denote measurement noise. Because a constant phase shift cannot affect the overall image, the wavefront sensor must detect local phase differences so that the term

$$\sum_{\ell} B_{j,\ell}\, S(\ell,t)$$

in Eq. 1 should describe local phase differences. Let us say that this term describes a local wavefront tilt or phase gradient at point j. Thus, the matrix (B) can be considered as a matrix describing the gradient operator (∇).

Now let us consider the wavefront corrector and data processor. The wavefront corrector produces a phase shift at a point, ℓ, denoted by $Z(\ell,t)$. The data processor, denoted by $L_{\ell,j}(t)$, operates on the wavefront sensor measurement, $P(j,t)$, to produce the wavefront corrector phase shift, $Z(\ell,t)$:

$$Z(\ell,t) = \sum_{j} \int_{-\infty}^{t} dt'\, L_{\ell,j}(t-t')\, P(j,t') \qquad (2)$$

The kernel (L) contains both the wavefront corrector impulse response (Green's function) and any filtering that may be required. If ϕ is the input wavefront, Eqs. 1 and 2 can be solved to yield:

$$S = [1 - LB]^{-1} [\phi + L(n + P_o)] \qquad (3)$$

An obvious condition on L is that $LP_o = 0$, which says that the system does not respond to an unperturbed input.

Optimum Performance

At this point, it is desirable to determine the form of L or the filtering required to obtain optimum system performance. Optimum system performance for a linear stochastic control system is known to be obtained by a least-square minimization of the expected estimation error. For an image compensation system, it is convenient to minimize the wavefront estimation error (E) defined by:

$$E = \sum_{\ell,\ell'} C(\ell, \ell') < S(\ell, t)\, S(\ell', t) > \qquad (4)$$

where C is a matrix that removes any contribution to E from an overall piston error. The averaging brackets, < >, in Eq. 4 denote two ensemble averages: (1) the average over all random wavefronts and (2) the average over all noise signals, n(j, t).

From Eq. 3, the expectation of the error can be written as:

$$<S_1 S_2> = \left[1 - L_1 B_1\right]^{-1} \left[1 - L_2 B_2\right]^{-1} \left[<\phi_1 \phi_2> + L_1 L_2 <n_1 n_2>\right] \qquad (5)$$

where the matrix products in Eq. 5 are recognized as convolutions. The first term in Eq. 5 can be interpreted as the servo lag error, and the second term as the noise error. From the theory of linear stochastic control, we know that the optimum filter (L) trades off tracking error and noise error so that their sum is a minimum. In fact, the optimum filter under these conditions has become known as a Wiener filter.(1-3) The classic Wiener filter in this instance suffers from one serious drawback -- it is not a causal filter, which it must be.

To obtain the optimum causal filter, one must solve what is known as the Wiener-Hopf equation. This problem has recently been formulated for image compensation systems by Dyson,(4) who indeed has found that a Wiener-Hopf equation has to be solved in order to find the optimum filter. The solution is:

$$L_{\ell,j}(t_\ell - t_j) = \sum_{\ell'} \frac{K_{\ell,\ell'}(t_\ell - t_{\ell'})\, B_{j,\ell'}}{N_o(j)}\, \delta(t_j - t_{\ell'}) \qquad (6)$$

or, in matrix notation,

$$L = K\tilde{B} N_o^{-1} \qquad (7)$$

with $\tilde{B}$ indicating the transpose of B and $N_o(j)$ the noise in channel j. The matrix (K) is a causal matrix whose physical significance is shown shortly; it is determined by the following integral equation:

$$K = \tilde{H} + D\tilde{B}N_o^{-1}\, B\tilde{H} - D \qquad (8)$$

where D is the space/time matrix of the atmospheric structure function and $\tilde{H}$ is an anticausal kernel satisfying the Wiener-Hopf equation:

$$\tilde{H} + P_a D\tilde{B}N_o^{-1}\, B\tilde{H} = P_a D \qquad (9)$$

The operator (P_a) is an anticausal operator such that:

$$P_a U(t, t') = U(t, t') \quad \text{for } t \le t'$$

$$P_a U(t, t') = 0 \quad \text{for } t > t'$$

The problem confronted by the system designer is that the structure function matrix (D) is not well known and will change from day to day and minute to minute. Thus, what the system designer expects to learn from Eqs. 8 and 9 are some general properties that the filter should possess. This point will be returned to in a minute, but as a passing observation, we note that if we substitute Eqs. 1 and 6 into Eq. 2, we have:

$$Z(\ell, t) = \sum_j \int_{-\infty}^{t} dt_{\ell'} K_{\ell,\ell'} (t - t_{\ell'}) \frac{B_{j,\ell'}}{N_o(j)} \left[\sum_{\ell''} B_{j,\ell''} S(\ell'', t_{\ell'}) + n(j, t_{\ell'})\right] \tag{10}$$

If we interpret

$$\frac{B_{j\ell'} \sum_{\ell''} B_{j,\ell''} S(\ell'', t_{\ell'})}{N_o(j)}$$

as a wavefront corrector force and K as the corrector Green's function, the Wiener-Hopf equation determines the optimum wavefront corrector. Since B ~ gradient operator and $B^2 \sim \nabla^2$,

$$\text{Force} \sim \nabla^2 S \tag{11}$$

Eq. 11 is the motivation for referring to the corrector as a rubber mirror because it is the equation of a membrane.

Structure Function Properties

To determine some properties of the causal filter, we turn now to an examination of the atmospheric statistics. The Wiener spectrum W(k, f) of the atmospheric fluctuations can be written as:

$$W(k, f) \sim k^{-11/3} \int_0^{\infty} dh\, C_N^2(h)\, \delta(k \cdot v_h - f) \tag{12}$$

where k is spatial frequency, f is temporal frequency, $C_N^2(h)$ is the atmospheric turbulence height profile, and v_h is the wind velocity at height, h. The following illustration is a plot of some recently measured profiles by Vernin and Roddier.[5] The measurements show a layered structure; this means that in an image compensation system, the time dependence of the wavefront measurements will appear to propagate across the aperture at various rates corresponding to the velocities associated with the various atmospheric turbulence layers.

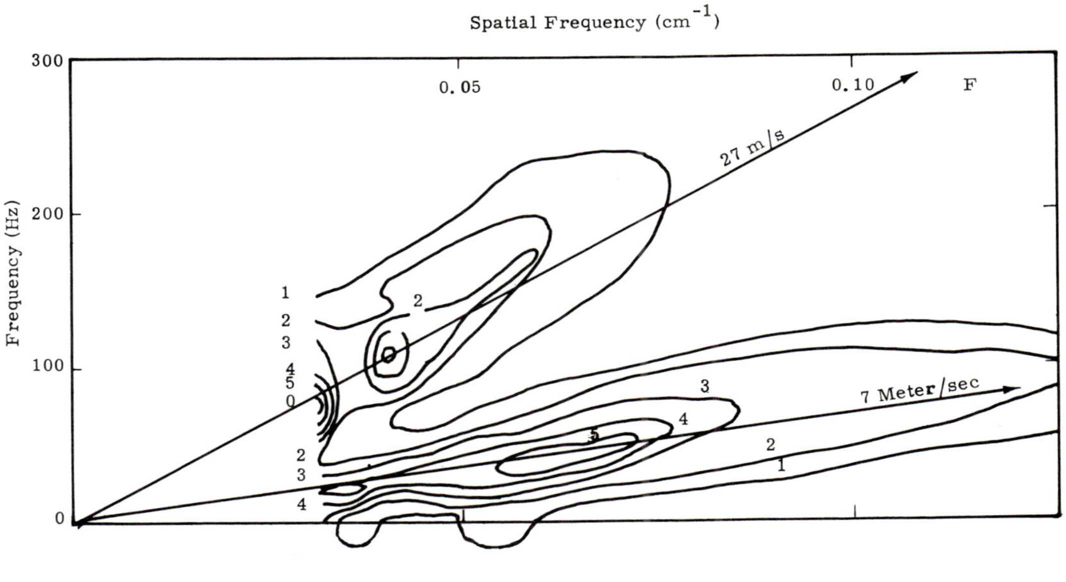

Contours are Lines of Constant C_N^2 Measured by Vernin and Roddier

A solution, neglecting edge effects, for the optimum filter (K) when the structure function is of the form $D \sim D(r)\,\delta(r-vt)$ also shows the form $\delta(r-vt)$, except that because of causality, only points (r) that lie upstream are allowed to contribute. Because edge effects have been neglected, this optimum filter essentially permits data gathering from infinitely far upstream; therefore, the filter bandwidth would be infinitely narrow so that the noise contribution to the error is zero. When edge effects are considered, the corrected wavefront error will be spatially nonstationary with aperture points upstream having much more error than aperture points downstream.

Based on the measurements of Vernin and Roddier, the optimum filter is not a single propagating filter, $\delta(r-vt)$, but a linear combination of propagating filters whose relative strengths are determined by the atmospheric profile, $C_N^2(h)$, through the Wiener-Hopf equation. Because the winds and the profile, $C_N^2(h)$, vary from time to time, the systems designer is concerned about the sensitivity of filter performance to such changes. Ultimately, this question of sensitivity must be answered by the actual operation of such a system in the field with the real atmosphere. Simulations performed at Perkin-Elmer give encouragement that the sensitivity may not be very critical.

Summary

Some general properties of a ccntrol system for compensating atmospherically distorted wavefronts in a telescope imaging system have been discussed. The optimum filter for such a system is obtained as a solution to the Wiener-Hopf equation. Because recent atmospheric measurements suggest a layered profile of the atmosphere, the Wiener-Hopf solution can be viewed as a propagating filter that moves data through the system aperture at a rate and direction related to the wind speed.

References

1. Wiener, N. "The Interpolation, Extrapolation, and Smoothing of Stationary Time Series", National Defense Research Committee; reprinted together with two papers by N. Levinson, John Wiley and Sons, Inc., New York (1949).

2. Kolmogoroff, A., "Interpolation and Extrapolation van Stationaren Zufalligen Folgen", Bull. Acad. Sc. (USSR), Ser. Math., 5, pp 3-14 (1941).

3. Bode, H. and Shannon, C., "A Simplified Derivation of Linear Least Square Smoothing and Prediction Theory", Proc. IRE, 38, 4, p. 417 (1950).

4. Dyson, F., "Photon Noise and Atmospheric Noise in Active Optical Systems", JOSA, 65, 5, p. 551 (1975).

5. Vernin, J. and Roddier, F., "Experimental Determination of Two-Dimensional Spatiotemporal Power Spectra of Stellar Light Scintillation. Evidence for a Multilayer Structure of the Air Turbulence in the Upper Troposphere", JOSA, 63, 3, p. 270 (1973).

Session 2
ATMOSPHERIC MEASUREMENTS

Session Chairman
J. W. Goodman
Department of Electrical Engineering, Stanford University

AN EXPERIMENT FOR MEASURING EFFECT OF ATMOSPHERIC TURBULENCE ON A VERTICAL OPTICAL PATH

Robert R. Shannon and W. Scott Smith
Optical Sciences Center, University of Arizona
Tucson, Arizona 85721

Abstract

The experiment to be described is intended to obtain a measure of the limit on the phase coherent propagation of light along a vertical path from a point source. This experiment utilizes an argon laser to generate a pair of coherent point sources in an aircraft. These sources generate a fringe pattern on the ground. As the aircraft flies along a track over a set of ground detectors, a frequency modulated signal is generated in the photoelectric detectors, the frequency modulation being proportional to the phase shift difference along the two paths from the plane. A Fourier transform of the data produces a measure of the phase coherence function or the Atmospheric Transfer Function of the atmospheric path. A modification of the direct experiment uses polarized sources in the transmitter to permit reconstruction of the phase difference function along the paths. From this data, the statistical parameters of the phase structure and log-amplitude behaviour can be obtained.

Introduction

The experiment to be described here is intended to measure the statistics of the upward propagation of a spherical wave through the atmosphere. A high altitude aircraft carries one half of the experimental equipment, while a ground station serves as the base station for the experiment. Previous experiments have used a coherent source on the ground and a shearing interferometer in the aircraft as shown in Fig. 1. The interferometer generates wavefront difference statistics as a function of time. In a practical sense, there are many difficulties in carrying out such an experiment. The experiment which we will describe uses a reverse situation in which a pair of coherent sources are located on the aircraft and a number of single point collectors are located at the ground station. This reversal maintains the geometry of the original propagation situation, and under this condition, reversability of the atmospheric path is obtained.

Theory

The theory of the experiment can be explained in two ways. On basis number one, we are endeavoring to obtain the partial coherence function of transmission through the atmosphere. If, as in Fig. 2, one imagines that the two sources on the aircraft generate a fringe pattern on the ground, this fringe pattern will be swept across a ground detector as the aircraft passes overhead. In the absence of any atmospheric errors (and of course vehicle induced errors) the fringe pattern will consist of a single frequency which is converted to a single temporal frequency by the scanning motion of the aircraft. Introduction of the atmosphere will introduce time varying fluctuations of phase difference between the two paths, thus introducing a time dependence or phase jitter in the position of the fringes. This is recorded as a frequency modulated signal by the ground equipment. The larger the amount of this frequency modulated phase shift, the lower will be the phase coherence function.

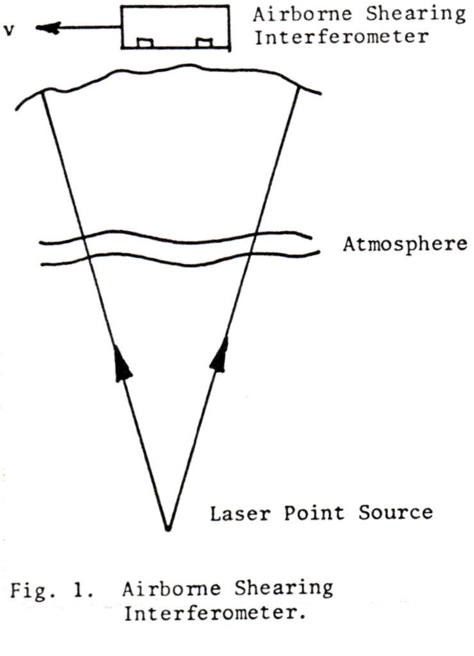

Fig. 1. Airborne Shearing Interferometer.

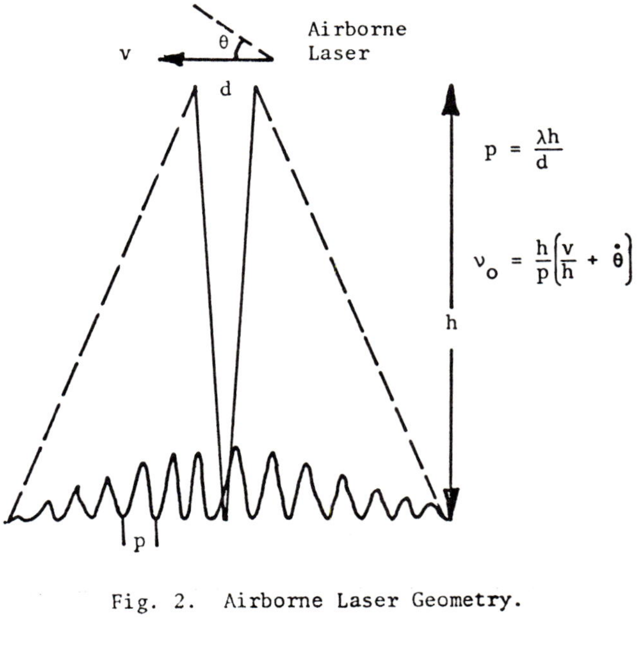

Fig. 2. Airborne Laser Geometry.

Using the hypothesis of a frozen atmosphere reasonable with the velocities of the aircraft involved here, we can write the atmospheric transfer function, or ATF, as

$$S(t) = A_1^2(t) + A_2^2(t) + 2A_1(t)A_2(t)\cos(2\pi wt + \Delta\phi) \tag{1}$$

where $A_{1,2}$ = Modulus of the wave amplitude received from point sources 1,2.

w = Fringe frequency (a constant)

$\Delta\phi$ = Atmospherically induced phase difference between the light received from point sources 1 and 2.

Eq. (1) describes the output of the receiver on the ground. The ATF can be directly related to the Fourier transform of the frequency modulated fringe function as detected on the ground,

$$ATF(\Delta x) = \frac{\int_0^T A_1A_2e^{i\Delta\phi}dt}{\frac{1}{2}\int_0^T \left(A_1^2 + A_2^2\right)dt} \tag{2}$$

where x = Point source separation (i.e. the shear)

$$ATF(\Delta x) = \frac{F.T\ \{s(t)\}\ \upsilon = w}{F.T\ \{s(t)\}\ \upsilon = o} \tag{3}$$

In this manner, the character of the experiment at a number of separations of sources produces a map of the ATF or phase coherence function of the wavefront as propagated between the sources at a single point on the ground. Having sufficient amounts of data will produce the average statistics to describe the behavior of the ATF.

A second basis for explaining the theory of the experiment is to recognise that what we are measuring is the phase difference between two points on the aircraft and one on the ground, just as in the original interferometer experiment. Obviously, by keeping a record of the phase shift as a function of time in the fringes during the aircraft's passage, a map of this phase difference as a function of time can be generated. An additional elaboration of the experiment involves the coding of the two sources on the aircraft into oppositely circularly polarized beams. The detectors on the ground then split these beams into two or more components, (in our case we chose four), and quadrature between the two sources is obtained. In this way, a more accurate record of the phase difference is obtained, since maintenance of the quadrature through four channels will allow piecing of the phase errors that occur in the presence of large log amplitude fluctuations of the intensity. There are, in fact, many complications in the data processing which takes place in terms of eliminating phase jumps and so on. In a fundamental manner, however, the calculation is straight forward.

There is, in addition, a problem brought about by the induced motion of the aircraft; in particular, pitch motion and instability of the package in the aircraft will produce a rocking of the two point sources and apparent change of phase of the function of time. It is necessary that this be removed from the data. There are two approaches available to accomplish this. The first approach uses an inertial sensor attached to the experimental package which gives a continuous recording of the pitch angle. This can be transmitted to the ground and recorded along with the time dependent signal from the optical detectors, and then used to generate the pseudo phase error which is subtracted from the calculated optical phase error. The second approach is to use a number of detectors and look for a common mode of pitch error on several detectors simultaneously. Both of these approaches have been included in this experiment. The latter approach has the advantage that any other disturbing factors, such as boundary layer turbulence, are eliminated by difference processing of the data. In addition, the use of a number of detectors arrayed in a line on the order of 400 to 500 feet on the ground greatly increases the amount of data that can be taken because of the increased probability that the aircraft will pass over a detector station on each pass.

Implementation

The implementation of the experiment is also reasonably straight forward, but with some subtleties. The aircraft used has sufficient power to supply the 10 kilowatts necessary to run an argon laser which produces 4 watts of output power. Initial investigation of the signal to noise level in the background as observed by the detectors indicated this level of light power is necessary to obtain an adequate signal to noise ratio. The light from the laser is split into two parts, the green 514.5nm light being used to provide information to a multiple channel experiment in which two circularly polarized beams, separated by about 1.9cm in the aircraft, are used as the source. The blue wavelength 488.0nm is used to provide light to a pair of point sources which are separable from 5 to about 50cm. Thus two correlated experiments can be carried out at the same time. Fig. 3 shows the diagram of the experimental package as placed in the aircraft. Fig. 4 shows the ground detectors used in the experiment. There are a total of 28 channels of data recorded; 10 single channel, 4 four channels, and 2 channels of tone encoded data from the inertial sensors in the aircraft. The data, when recorded, are examined by a computer. The computer selects one of the active detectors to provide an on-line indication of the signal being received. During the course of the experiment it is possible to monitor fringe contrast, ATF, and variation of the average frequency with time or pitch behavior in the aircraft.

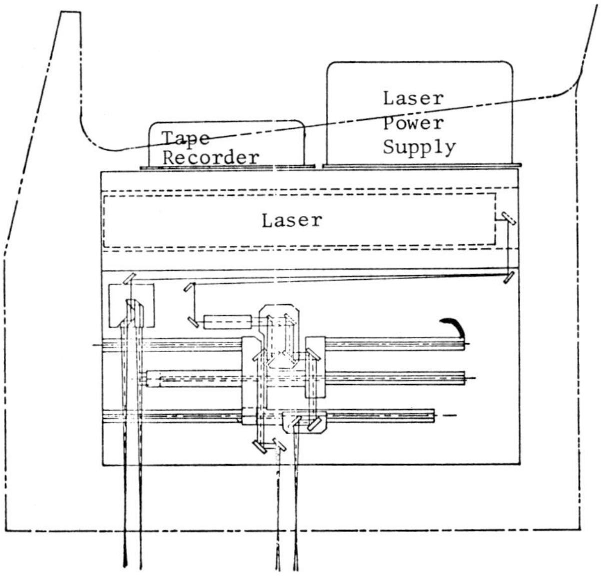

Fig. 3. Experimental package side view.

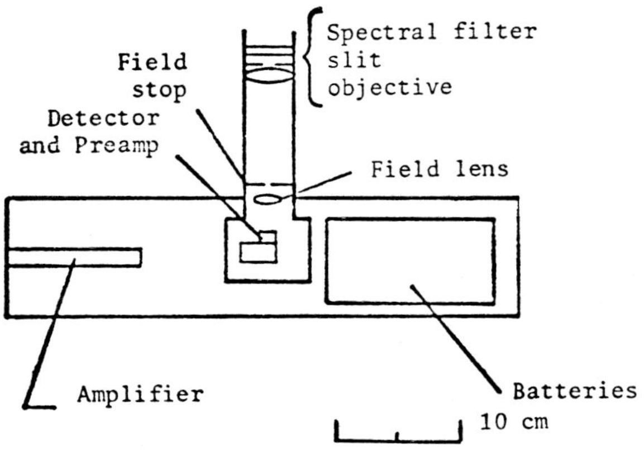

Fig. 4. Ground detector.

Since the aircraft being used is a very stable vehicle, it was decided to use passive isolators to eliminate high frequency vibration which led to the fact that our package, while very quiet, does have approximately a 5-Hz. resonant swing in the pitch direction. It has turned out that this error is very useful in the fact that it does show up clearly in much of the data being taken and is a good index of the magnitude of the pitch error. This of course must be removed in the final data.

Processing of the data consists of generating streams of wavefront phase data, and from this calculating the power spectrum of the phase fluctuations and phase difference as a function of time. These may then be corrected to the power spectrum of the phase as a function of time. There are obviously many numerical and mathematical subtleties to the handling of this data, which are too lengthy to dwell on here. Samples of data takes are shown in Figs. 5, 6, and 7. Figure 5 shows raw data from a single-channel detector. Each of the channels of a multi-channel detector have similar output with a different frequency resulting from the narrower separation. Figure 6 is a four-channel phase difference plot. The 5-Hz. pitch angle resonance of the passive isolators is evident. Figure 7 is a phase power spectrum generated from one of the four-channel data takes.

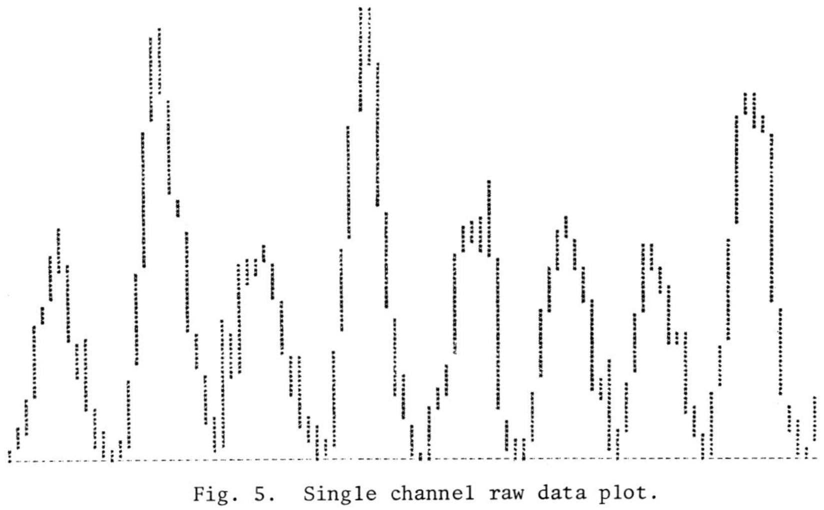

Fig. 5. Single channel raw data plot.

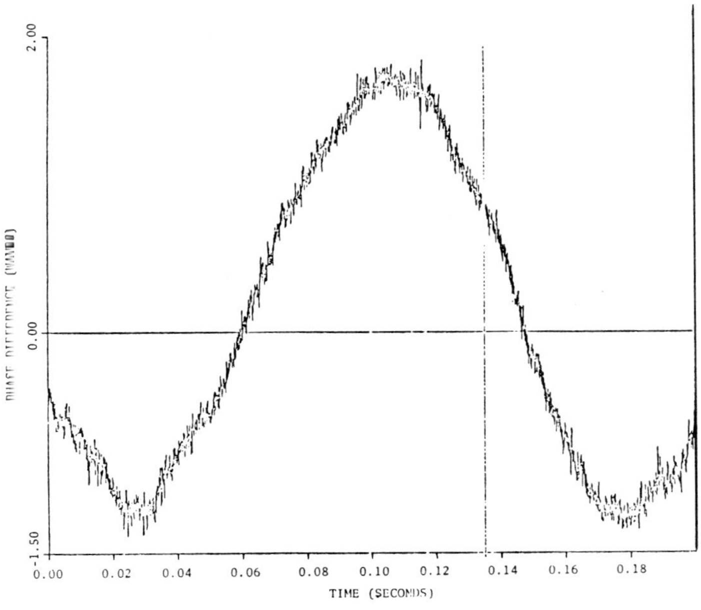

Fig. 6. Four-channel phase difference plot.

In the results that have been obtained from single-channel data is some statistical variations of the ATF, or phase coherence function that have been obtained from the single-channel data. Needless to say, an experiment of this sort is a very difficult one to carry through. The experiment is presently in progress and data is being generated. So far, experiments have been run only in Tucson, and primarily for the purpose of testing the experiment rather than the atmosphere. Preliminary indications are that the atmospheric effects are more benign than previous experiments have indicated.

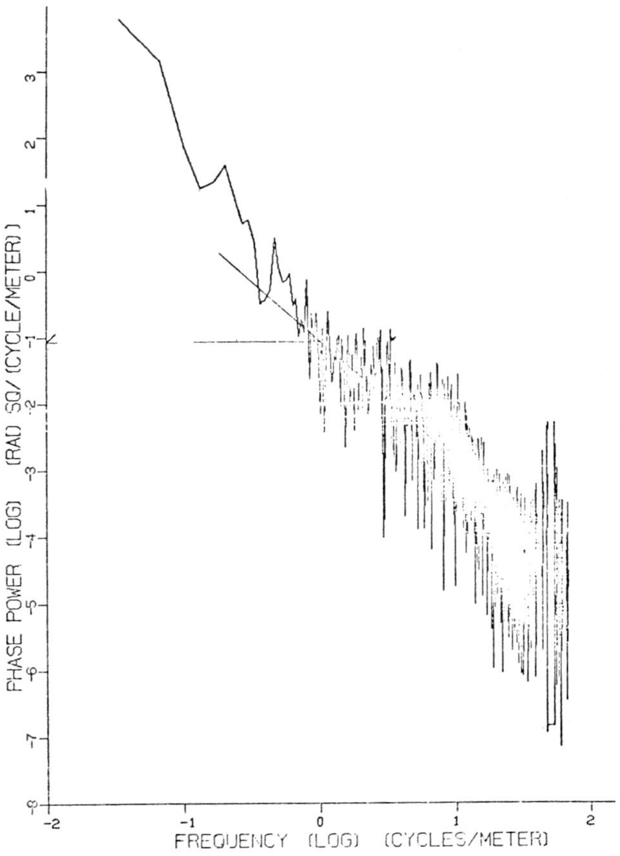

Fig. 7. Four-channel phase power spectrum.

STELLAR-SCINTILLATION MEASUREMENT OF THE VERTICAL PROFILE OF REFRACTIVE-INDEX TURBULENCE IN THE ATMOSPHERE

G. R. Ochs, R. S. Lawrence and T. Wang
National Oceanic and Atmospheric Administration
Environmental Research Laboratories
Boulder, Colorado 80302

and

P. Zieske
AVCO Everett Research Laboratory
Puunene, Hawaii 96784

Abstract

We observe the turbulence-induced scintillation (twinkling) of a single star to measure, from the ground, the vertical profile of refractive-index turbulence in the atmosphere. A linear combination, with appropriate weights, of the strength of the scintillations observed with receivers of various spatial wavelengths allows us to synthesize a path weighting function centered at a specific height. The central height and height resolution of the measurement can be controlled by changing the relative coefficients and spatial wavelengths of the receiver outputs. Twenty-minute measurements, made with stars of second magnitude or brighter and with a 36-cm Schmidt-Cassegrain telescope, show that the atmospheric turbulence can be divided into four independent height regions with reasonable accuracy. The measurements of the different spatial wavelengths are made sequentially with the same telescope, and the statistical stationarity of the atmosphere during the 20-minute observation period is crucial to the accuracy of the deduced profile. Stationarity is roughly checked by continuously monitoring the whole-aperture scintillation of the star. The observed profiles agree with the strength and general shape of accepted models of the atmosphere and with profiles obtained from aircraft-mounted and previously used balloon-borne in-situ sensors.

Introduction

For centuries, it has been known that observations of astronomical sources, particularly the twinkling of stars and the motion of stellar images, yield information about the turbulence in the atmosphere.[1-12] Scintillation measurements have usually been used by astronomers for site-selection purposes but, in recent years, some attention has been given to the problem of deducing the vertical profile of refractive-index turbulence from ground-based measurements of stellar scintillation.[8-12] The application of these developments to the design of an optical instrument for measuring refractive-index turbulence will permit the gathering of synoptic data relevant to problems of ground-based imaging of astronomical and orbital objects.

We have previously discussed[10] the feasibility of using either a single-star or a double-star pair as the source for remote probing of the vertical atmospheric turbulence profile. The advantage of using a double star as a source is that a relatively good height resolution would be obtained by use of the crossed-path technique.[11] However, there are few useful double stars in the sky, so it seemed to us worthwhile to build a single-star instrument although the altitude-resolution may not be so good as could be obtained from the crossed-path method. In a previous paper,[12] we discussed a technique for measuring the atmospheric turbulence profile aloft by using a single star as a source and a spatially-filtered detector array as a receiver. Here, we present some quantitative results in the form of measured vertical profiles of refractive-index turbulence at Boulder, Colorado and at Mt. Haleakala, Hawaii. A comparison of the ground-based measurements and airplane in-situ measurements is also presented.

Theory and Implementation

The general philosophy of our approach is to view the spatial structure of the stellar scintillation pattern through filters that pass only a narrow band of spatial frequencies. By observing the signal intensity through filters of different spatial wavelengths and combining the outputs with appropriate weights, we obtain a set of reasonably sharp-peaked path-weighting functions centered at different heights, and so measure the vertical turbulence profile.

If we assume that the turbulence is described by a Kolmogorov spectrum, then, from Eq. (12) of reference 12, the composite path-weighting function resulting from a combination of N filters is given by

$$W_c(z) = \sum_{i=1}^{N} R_i \; K_i^{8/3} \; W_i'(z,d_i) \; , \qquad (1)$$

where z is the height above the ground, N is the number of different wavelength filters used, R_i is the appropriate weight of the ith spatial wavenumber $K_i = 2\pi/d_i$, and d_i is the ith spatial wavelength. The W_i' in Eq. (1) are the weighting functions of the corresponding spatial wavelengths d_i given by Eqs. (9) to (11) of reference 12. In calculating the W_i', we have included the effect of a finite-sized array and the effect of a broad-band source (see ref. 12 for details). The number N of filters used is arbitrary. The larger N will give theoretically sharper weighting functions but the results will be more sensitive to experimental error and to non-stationarity of the atmosphere. In order to tolerate the non-stationarity and noise found in real observations, we restrict N to 3 or less. By trial and error combinations of the values of d_i and R_i, we obtained, with the help of a computer, a family of seven weighting functions peaked at different heights ranging from 2.25 km to 14.5 km and above (see Fig. 1). The peak altitude, the three spatial wavelengths needed for each linear combinations and the relative weights R_2 and $R_3 (R_1 \equiv 1)$ are shown in Table 1. The seven profile values are not independent. They can roughly be divided into four nearly independent layers (e.g., the odd-numbered combinations).

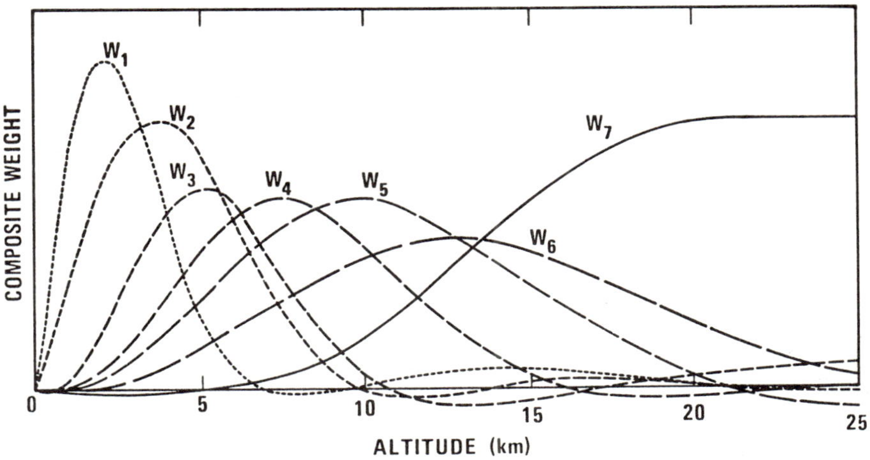

Fig. 1. Composite path-weighting functions obtained by linearly combining the weighting functions of three different spatial wavelengths (see text and Table 1 for details.)

Table 1. The peak altitudes and relative inverse areas (see Eq. (14)) of each of the seven weighting functions W_i shown in Fig. 1. The d_i are the spatial wavelengths that must be combined to produce each W, the relative weights being R_2 and $R_3 (R_1 \equiv 1)$.

	W_1	W_2	W_3	W_4	W_5	W_6	W_7
Peak (km) altitude	2.25	3.75	5.25	7.5	9.75	12.75	>14.5
$B \times 10^{14}$	2.39	1.91	2.57	1.84	1.38	1.46	0.87
d_1(cm)	5	6.5	7	8.5	10	11.5	14
d_2(cm)	8	9	5	6	6.5	7.5	9
d_3(cm)	15	15	11	13	15	15	7
R_2	- .65	- .6	- .38	- .35	- .27	- .24	- .73
R_3	- .33	- .36	- .53	- .56	- .66	- .67	+ .23

The C_n^2 of each profile value is given by

$$C_n^2 = B \sum_{i=1}^{N} R_i K_i^{8/3} \sigma_I^2(d_i), \qquad (2)$$

where

$$B = 1.87\times10^{-13} [\int_0^\infty dz\ W_c\ (z)]^{-1} \quad . \qquad (3)$$

In Eq. (2), $\sigma_I^2(d_i)$ is the measured irradiance variance of spatial wavelength d_i, C_n^2 is the refractive-index structure constant in $m^{-2/3}$, K_i is the spatial wavenumber in cm^{-1} and z is in km. By measuring the areas under the curves shown in Fig. 1 (for W_7, we cut it off at 25 km), we obtain the calibration factor B shown for each layer in the second line of Table 1.

We have developed a system that uses the spatial filtering technique to make an on-line computation of the refactive-index profile. An instrument package attached to a 36 cm Schmidt-Cassegrain telescope (Fig. 2) sequentially measures the variance of the intensity scintillation at different spatial wavelengths. The instrument package supplies to the analog-to-digital minicomputer four signal voltages proportional to the following:

Fig. 2. The telescope and the attached instrument box.

- $\sigma_I(d)$, the standard deviation of the spatially filtered signal for a specific spatial wavelength d,
- d , the wavelength of the spatial filter,
- σ_I , the log-irradiance standard deviation of the whole telescope aperture,
- I , the irradiance of the whole telescope aperture.

From the first two inputs we calculate a C_n^2 profile according to Eq. (2). The fourth input (I) is used to set signal levels, monitor the guiding of the telescope, and provide a record if clouds obscure the star. The log-irradiance standard deviation of the whole aperture is a weighted measurement of the refractive-index turbulence which approximately covers the region over which the instrument profiles C_n^2. While in principle this measurement is not needed to compute C_n^2 profiles, we have found it useful to compensate partially for the non-stationarity of the atmosphere during the 20-min measurement cycle. (See section IV of Ref. 12 for details of the design of the instrument.)

Results

The instrument is designed for use with stars of second magnitude or brighter. In order to guarantee enough signal-to-noise ratio (SNR), we first pointed the telescope to a zero magnitude star, Vega. The measurement was taken at Table Mountain, a flat mesa located 12 km north of Boulder, Colorado. Some measurements of the typical C_n^2 vertical profiles are shown in Fig. 3. The "altitude" shown is actually slant range from the observatory 1.7 km above sea-level. Zenith-angle corrections, amounting to less than 2 percent, have been neglected. Variations from night to night are large, extending over a range of one hundred to one. Our measurements show general agreement with Hufnagel's model.(13) Since we are in the lee of the Rocky Mountains where mountain-wave effects are common, we expect to see large night-to-night variations of the vertical C_n^2 profile. A typical variation over a period of two and one-half hours in a single evening is shown in Fig. 4. We see that there are noticeable changes in a couple of hours.

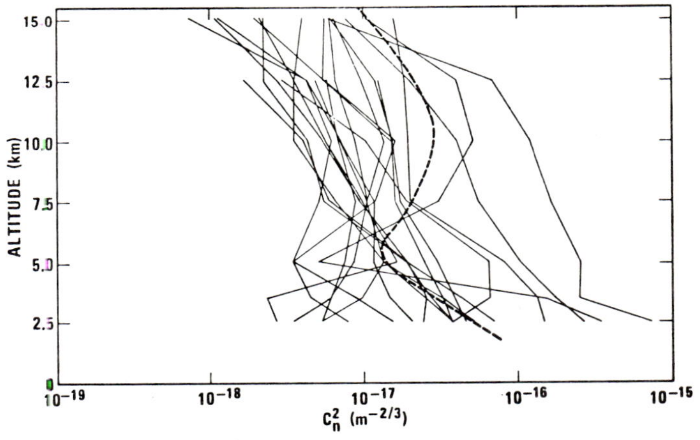

Fig. 3. Typical optically measured C_n^2 vertical profiles at Boulder, Colorado. Each curve is the average of the measurements of a single evening. The dashed line represents the statistical average profile of Hufnagel's latest model.(13)

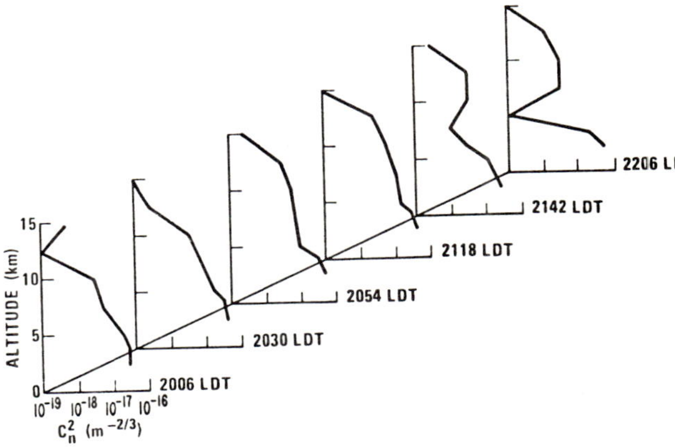

Fig. 4. Typical variation of the vertical C_n^2 profile in a single evening.

The instrument was then moved to Maui, Hawaii and installed at the ARPA Maui Optical Station (AMOS), an observatory located on the top of Mt. Haleakala, 3 km above sea level. The night-to-night variations are shown in Fig. 5. Again the "altitude" is range from the observatory. Zenith-angle corrections, always less than 10 percent, have been neglected. Hour-to-hour variations in a single evening are shown in Fig. 6. Both Figs. 5 and 6 show variations similar in magnitude to those in the data from Colorado. However, the mean value of C_n^2 is smaller than in the Colorado data. This result might be expected since the AMOS observatory is on an isolated peak 3 km above sea level whereas the Colorado measurements were taken at the level of the surroundings and in the lee of 2000 km of mountainous terrain, and since the winds aloft (and therefore, presumably, the wind shear) are weaker in the tropics.

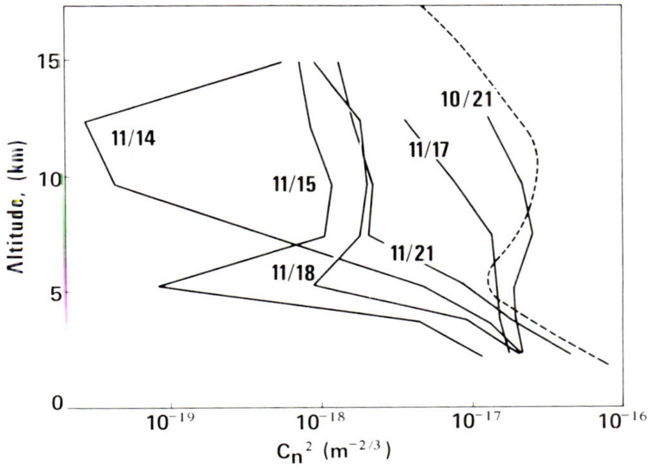

Fig. 5. Typical optically measured C_n^2 vertical profiles at Mt. Haleakala, Hawaii. Each curve is the average of the measurements of a single evening. The dashed line represents the statistical average profile of Hufnagel's latest model.(13)

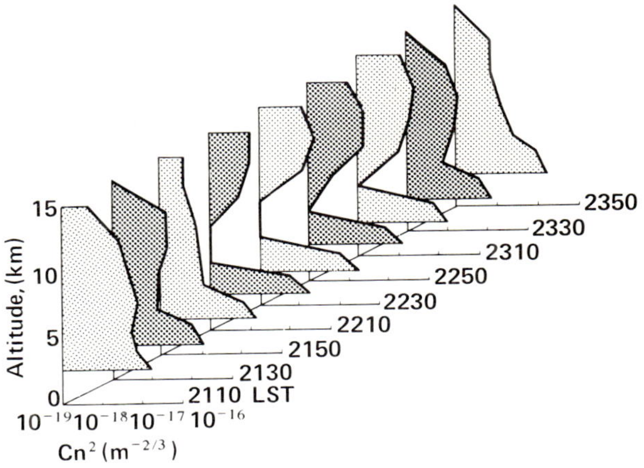

Fig. 6. Typical variation of the vertical C_n^2 profile in a single evening.

It is difficult to obtain an independent confirmation of our C_n^2 measurements. Rawinsonde data are available but such wind and temperature information does not have sufficient detail to make estimates of C_n^2. It is possible, however, to derive C_n^2 from direct small-scale measurements of the temperature field, and we made measurements of this type from a light aircraft over Haleakala. Only the lowest data point obtained with the optical instrument could be compared because of aircraft altitude limitations.

We derive C_n^2 by assuming that, over small scale sizes, air density and therefore C_n^2 are affected primarily by temperature irregularities, and very little by variable pressure and composition. At optical wavelengths,

$$C_n^2 = \left(\frac{79P}{T^2} 10^{-6}\right)^2 C_T^2 \quad , \tag{4}$$

where P is the pressure in millibars, T is the Kelvin temperature, and C_T^2 is the temperature structure parameter. Assuming that the spectrum of turbulence follows the Kolmogorov law, a single sensor mounted on an aircraft traveling at velocity v, and arranged to respond to angular frequencies from ω_L to ω_H, may be used to measure C_T^2. It can be shown that[14]

$$C_T^2 = \frac{2.68[\langle T^2 \rangle_{AV} - \bar{N}^2]}{(v/\omega_L)^{2/3} - (v/\omega_H)^{2/3}} \quad , \tag{5}$$

where $\langle T^2 \rangle_{AV}$ is the mean-square temperature fluctuation, and $\bar{N}$ is the noise from all sources, principally the aircraft engine and propellor.

The sensor is mounted on the wing of the aircraft, and consists of a platinum wire 6μm in diameter, wound on an open square form 0.35 cm on a side and 2 cm long. The low-frequency cutoff is ω_L = 1.45 sec^{-1} (0.23 Hz), and the high-frequency limit ω_H is 2070 sec^{-1} (330 Hz). A detailed description of the measurement of C_n^2 with this temperature probe can be found in reference 14.

Profile measurements were made during spiral descents over Haleakala, as there is less background noise at reduced engine power. In Fig. 7, we compare aircraft measurements and optical measurements made during the evening hours of November 17, 18, and 21, 1975 over Mt. Haleakala.

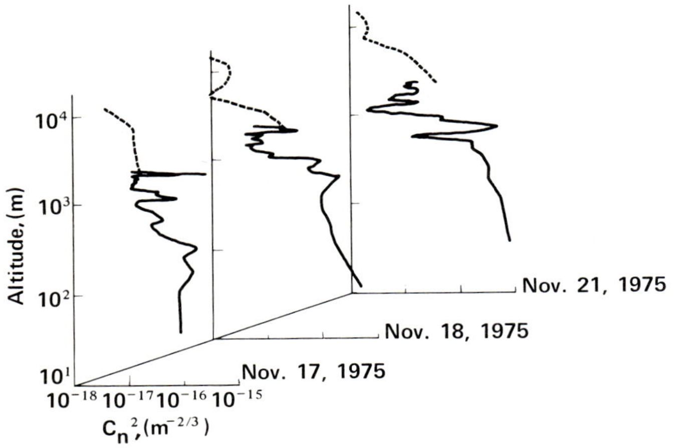

Fig. 7. Comparison of aircraft in-situ measurements (solid line) and optical measurements (dotted line).

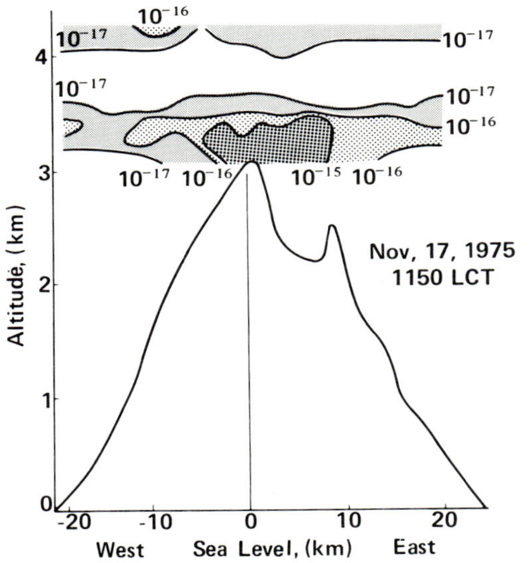

Fig. 8. Two-dimensional C_n^2 contours above Mt. Haleakala, Hawaii during daytime.

There are several difficulties with such a comparison. First, the spatial average of the lowest optical measurement covers about 3 km in height whereas the aircraft measurement is averaged over a height range of only 75 m. Also, the measurements do not exactly correspond in time or space. The horizontal variation of C_n^2 can be seen in

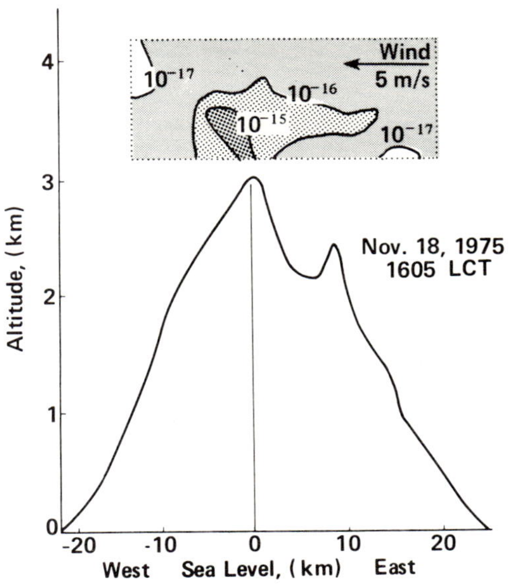

Fig. 9. Two-dimensional C_n^2 contours above Mt. Haleakala, Hawaii during daytime.

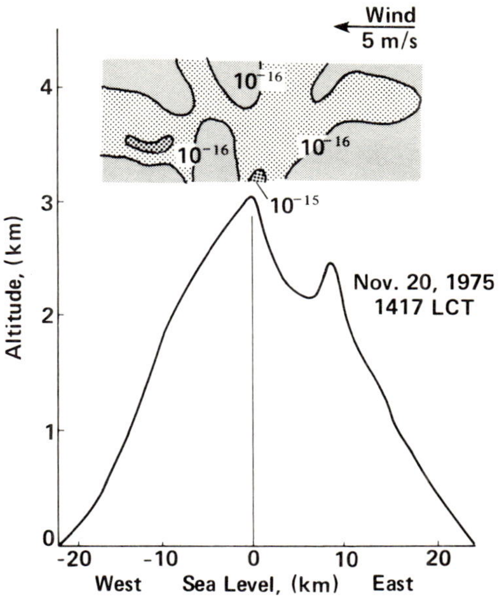

Fig. 10. Two-dimensional C_n^2 contours above Mt. Haleakala, Hawaii during daytime.

Figs. 8 - 10, obtained from horizontal aircraft flights along the direction of the prevailing wind. The horizontal separation of the aircraft and optical meaurements was perhaps a kilometer or more, and one can see that a considerable variation might be expected from such a separation. Allowing for these factors, we feel that the agreement is quite satisfactory.

Conclusions

We have demonstrated that the vertical C_n^2 profile can be measured by observing the turbulence-induced optical scintillation of a single star with a ground-based, spatially-filtered telescope. The whole atmosphere can be divided into four nearly independent layers. There is good agreement between the optical remote measurements and aircraft in-situ measurements. Night-to-night and hour-to-hour variations have been observed in the measured C_n^2 profiles taken at Boulder, Colorado, and on top of Mt. Haleakala in Hawaii. This is a potentially useful device to measure the vertical turbulence profile provided that a sharp altitude resolution is not required.

The primary support for this project was provided by RADC(OCSE), Griffiss AFB, New York 13441 under contract F30602-74-F-0108. The Scientific Monitor was Raymond P. Urtz, Jr.

References

1. Douglass, A.E., "Atmosphere, Telescope and Observer" Popular Astronomy, June, 1897 (Reprinted in Amateur Telescope Making, Book Two, pp. 585-605, A. G. Ingalls, ed. Scientific American, Inc. (1957)).

2. Pickering, W. H., "The Shadow Bands," Popular Astronomy, 33:, 1-2, 1925.

3. Gifford, F., and Mikesell, A. H., "Atmospheric Turbulence and the Scintillation of Starlight,", Weather 8:7, 195-197, 1953.

4. Protheroe, W. M.,"The Motion and Structure of Stellar Shadow-band Patterns," Quart. J. Roy. Met. Soc. 90:388, 27-42, 1964.

5. Townsend, A. A.,"The Interpretation of Stellar Shadow-bands as a Consequence of Turbulent Mixing," Quart. J. Roy. Met. Soc 91:387, 1-9, 1965.

6. Vernin, J. and Roddier, J.,"Experimental Determination of Two-Dimensional Spatio-Temporal Power Spectra of Stellar Light Scintillation. Evidence for a Multilayer Structure of the Air Turbulence in the Upper Troposphere," J. Opt. Soc. Am. 63:3, 270-274, 1973.

7. Roddier, C. and Roddier, F.,"Correlation Measurements on the Complex Amplitude of Stellar Plane Waves Perturbed by Atmospheric Turbulence," J. Opt. Soc. Am. 63:6, 661-663, 1973.

8. Rocca, A., Roddier, F., and Vernin, J.,"Detection of Atmospheric Turbulent Layers by Spatiotemporal and Spatioangular Correlation Measurements of Stellar-Light Scintillation," J. Opt. Soc. Am. 64:7, 1000-1004, 1974.

9. Vernin, J. and Roddier, F, "Détection au Sol de la Turbulence Stratospherique par Intercorrélation Spatioangulaire de la Scintillation Stellaire," Comptes Rendus Acad. Sci. Paris (French), 280:B, 463-465, 1975.

10. Ochs, G. R., S. F. Clifford, R. S. Lawrence, and Ting-i Wang, "Development of a Ground-Based Optical Method for Measuring Atmospheric Turbulence Aloft", NOAA Tech. Rept. ERL297-WPL30, Superintendent of Documents, U.S. Gov't Printing Office, Washington, D. C. 20402, 1974.

11 Wang, Ting-i, S. F. Clifford, and G. R. Ochs,"Wind and Refractive-Turbulence Sensing Using Crossed Laser Beams," Appl. Opt 13:11, 2602-2608, 1974.

12. Ochs, G. R., Ting-i Wang, R. S. Lawrence, and S. F. Clifford, "Refractive-Turbulence Profiles Measured by One-Dimensional Spatial Filtering of Scintillations, " (submitted to Applied Optics, 1976).

13. Hufnagel, R. E.,Variations of Atmospheric Turbulence, Optical Society of America Topical Meeting on Optical Propagation Through Turbulence, Boulder, Colorado July 1974.

14. Lawrence, R. S., G. R. Ochs, and S. F. Clifford,"Measurement of Atmospheric Turbulence Relevant to Optical Propagation," J. Opt. Soc. Am. 60:6, 826-830, 1970.

A NEW TURBULENCE SENSOR USING ATMOSPHERIC DISPERSION

Richard H. Hudgin
Itek Corporation
Lexington, Massachusetts 02173

Abstract

A new turbulence sensor has been designed which uses the dispersion in the atmosphere to intersect two beams of narrow band light at different altitudes. The wavefront tilts of the two beams are correlated which gives a weighted average of the turbulence distribution in the atmosphere. By varying the separation of the two beams at ground level, the altitude at which they intersect is shifted from ground level to infinity. The correlation of tilt for several separations allows the distribution of turbulence to be deduced with fair speed and accuracy.

Introduction

An atmospheric turbulence sensor has been invented for measuring the turbulence distribution. The fundamental concept of the turbulence sensor is to use two intersecting beams of light through the atmosphere and to correlate the tilts of the two beams. This time averaged correlation is a weighted average of the turbulence in the atmosphere and is weighted most heavily where the two beams intersect. By changing the altitude at which the beams intersect, different weightings can be achieved and then unscrambled to give the turbulence strength as a function of altitude.

Dispersion Shift

The intersecting beams are achieved in practice by using the chromatic aberration of the atmosphere. The air acts like a dispersive prism of exponentially varying refractive index. The relative displacement versus altitude as a function of wavelength λ is very nearly given by

$$b(\lambda,h) \approx s(\lambda)e^{-h/h_o} \qquad (1)$$

where h_o = 8300 meters is the scale height of the atmosphere.[1]

$S(\lambda)$ is related to the refractive index of the air $n(\lambda)$ by

$$\frac{S(\lambda)}{h_o} = (n(\lambda)-1)\tan\Theta_o\sec\Theta_o \qquad (2)$$

where Θ_o is the zenith angle of the light beam which in practice comes from a convenient star. (Derived from $n(\lambda,h) \times \sin\Theta(h)$ = constant.)

The refractive index of air is given in the American Institute of Physics Handbook page 6-111 as[2]

$$n(\lambda) = 1 + 6.4318\text{x}10^{-5} + \frac{2.949330\text{x}10^{-2}}{146 - \sigma^2} + \frac{2.5536\text{x}10^{-4}}{41 - \sigma^2} \qquad (3)$$

where σ is $1/\lambda$ in microns^{-1}. $n(\lambda)-1$ must be corrected for the altitude of the observatory by multiplying $n(\lambda)-1$ by the site pressure/sea level pressure.

Working between 400 and 600 nanometers, which the calculation says is effective, gives a change in $n(\lambda)$ of

$$n(600) - n(400) = 1.0001994 - 1.0002036 = -4.2\text{x}10^{-6} \qquad (4)$$

After going through the entire atmosphere, this corresponds at 60° zenith angle to a difference in displacement of

$$S(600) - S(400) = h_o\left(n(600) - n(400)\right) \tan\Theta_o\sec\Theta_o = 12.1 \text{ centimeters} \qquad (5)$$

Thus at a separation of 12.1 centimeters the 600 and 400 nanometer beams are coincident in the uppermost reaches of the atmosphere, and at zero separation they are coincident only at the lowermost part of the atmosphere. Wherever the beams are coincident, the tilts caused by the nearby turbulence cause the most correlation between the total tilts of the 400 and 600 nanometer beams. By correlating the tilts using different spacings, the turbulence is weighted in different ways and the distribution of turbulence can be deduced.

Tilt Correlations

Let $\phi_a(\vec{x},h)dh$ be the optical path lag caused by passing through a layer of thickness dh at height h. Then the wavefront for light of wavelength λ passing through the entire atmosphere is

$$\phi(\vec{x}) = \int dh\, \phi_a[x + S(\lambda)\,(e^{-h/h_o} - 1), h] \tag{6}$$

If this wavefront is imaged onto a detector with an L_x by L_y aperture, then the average tilt is

$$\overline{\frac{\partial\phi}{\partial x}} = \frac{1}{L_x L_y} \int_0^{L_x} dx \int_0^{L_y} dy \int_0^{\infty} dh \frac{\partial\phi_a}{\partial x} [\vec{x} + S(\lambda)(e^{-h/h_o} - 1), h] \tag{7}$$

In reality we are passing a wavelength band from λ' to λ'' with a relative photoelectron intensity of $I(\lambda)$ [due to atmospheric transmission, black body spectrums, and phosphor response.] Thus the slope must be averaged over the entire bandwidth.

$$\overline{\frac{\partial\phi}{\partial x}} = \int_{\lambda}^{\lambda''} d\lambda \int_0^{L_x} dx \int_0^{L_y} dy \int_0^{\infty} dh \frac{\partial\phi_a}{\partial x} [\vec{x} + S(\lambda)\,(e^{-h/h_o} - 1), h] I(\lambda) \tag{8}$$

$$\times \left[L_x L_y \int_{\lambda'}^{\lambda''} d\lambda\, I(\lambda) \right]^{-1}$$

Now consider two wavelength bands, one centered around 400 nanometers and the other around 600 nanometers. Where the apertures are shifted one relative to the other by an amount δ. Then the correlation between tilts is

$$\left\langle \frac{\partial\phi_1}{\partial x} \frac{\partial\phi_2}{\partial x} \right\rangle = \int_{\lambda_1'}^{\lambda_1''} d\lambda \int_{\lambda_2'}^{\lambda_2''} d\bar{\lambda} \int_0^{L_x} dx \int_0^{L_x} d\bar{x} \int_0^{L_y} dy \int_0^{L_y} d\bar{y} \int_0^{\infty} dh \int_0^{\infty} d\bar{h} \tag{9}$$

$$\times \left[L_x^2 L_y^2 \int_{\lambda_1'}^{\lambda_1''} d\lambda\, I(\lambda) \int_{\lambda_2'}^{\lambda_2''} d\bar{\lambda}\, I(\bar{\lambda}) \right]^{-1} I(\lambda) I(\bar{\lambda}) \sec\Theta_o$$

$$\times \left\langle \frac{\partial\phi_a}{\partial x} [\vec{x} + S(\lambda) e^{-h/h_o}, h] \frac{\partial\phi_a}{\partial x} [\vec{\bar{x}} + \delta + S(\bar{\lambda}) e^{-\bar{h}/h_o}, \bar{h}] \right\rangle$$

Now the statistics for the atmospheric turbulence are assumed to be Kolmogoroff which means that we are assuming zero inner scale and infinite outer scale. In this case, the phase slope correlation is[3]

$$\left\langle \frac{\partial\phi_a}{\partial x} (\vec{x},h) \frac{\partial\phi_a}{\partial x} (\vec{\bar{x}},\bar{h}) \right\rangle = \delta(h-\bar{h})\; 2.91\; C_n^2(h) \frac{\partial^2}{\partial x^2} |\vec{x} - \vec{\bar{x}}|^{5/3} \tag{10}$$

where $C_n^2(h)$ = turbulence strength as a function of altitude. Note: Units of $\frac{\partial\phi_a}{\partial x}$ are radians.

The $\frac{\partial^2}{\partial x^2}$ and $\delta(h-\bar{h})$ allow three of the eight integrations to be done easily to give

$$\left\langle \frac{\partial\phi_1}{\partial x}\frac{\partial\phi_2}{\partial x}\right\rangle = \int_{\lambda_1'}^{\lambda_1''} d\lambda \int_{\lambda_2'}^{\lambda_2''} d\bar{\lambda} \int_0^{L_y} dy \int_0^{L_y} d\bar{y} \int_0^{\infty} dh \; C_n^2(h) \qquad I(\lambda)I(\bar{\lambda})\sec\Theta_o \tag{11}$$

$$\times \left[L_x^2 L_y^2 \int_{\lambda_1'}^{\lambda_1''} d\lambda \int_{\lambda_2'}^{\lambda_2''} d\bar{\lambda} \; I(\lambda)I(\bar{\lambda}) \right]^{-1} W(\delta,h,y-y')$$

where

$$W(\delta,h,y-y') = \left[\begin{array}{l} [(y-\bar{y})^2 + (L_x + \delta + (S(\bar{\lambda})-S(\lambda))e^{-h/h_o})^2]^{5/6} \\ -2\,[(y-\bar{y})^2 + (\delta + (S(\bar{\lambda}) - S(\lambda))e^{-h/h_o})^2]^{5/6} \\ +\,[(y-\bar{y})^2 + (-L_x+\delta+(S(\lambda)-S(\lambda))e^{-h/h_o})^2]^{5/6} \end{array} \right]$$

and Θ_o is the zenith angle which increases path length by $\sec\Theta_o$.

Turbulence Weighting Function

The function W weighting the turbulence is plotted for λ = 400 nanometers, $\bar{\lambda}$ = 600 nanometers, δ = 6 centimeters in figure 1 assuming $L_x = L_y = 0$. The actual weighting function will not be as sharp since it is averaged over wavelength bands, but the general behavior will be the same away from the singularity.

Equation 11 has been averaged over various subapertures and various wavelength bands around 400 and 600 nanometers using the photoelectron intensity curves of an S-20 cathode. Various separations δ have been chosen so that the weighting function peaks in the center of the atmospheric layers whose turbulence is being sought. This means that if 6 layers of atmosphere are to be measured between 0 and 12000 meters, then δ's will be chosen which peak the weighting function at 1000, 3000, 5000, 7000, 9000, and 11000 meters. Then these six correlations are processed to give the strength of turbulence in each layer.

Error Analysis

To see how this is done and to get an error analysis of the resulting calculation, let δ_i be the ith shift ($i=1\rightarrow N$ for N measurements used to deduce N layers of turbulence), C_i be the ith correlation, and W_{ij} be the weight of the jth layer turbulence in the C_i measurement. i.e.

$$C_i = \sum_{j=1}^{N} W_{ij}\; C_n^2(h_j) \tag{12}$$

where $C_n^2(h_i)$ = mean $C_n^2(h)$ averaged over the ith layer of atmosphere.

then

$$W_{ij} = \int_{\lambda_1'}^{\lambda_1''} d\lambda \int_{\lambda_2'}^{\lambda_2''} d\bar{\lambda} \int_0^{L_y} dy \int_0^{L_y} dy' \int_{H(j-1)/N}^{H\,j/N} dh \quad C_n^2(h) \qquad I(\lambda)I(\lambda') \tag{13}$$

$$\times W(\delta_i,h,y-y')$$

Now the actual hardware will be sampling the correlation and will not give C_i perfectly due to random statistics. Let the error be N_i for measurement i. Solving for the $C_n^2(h_j)$ estimates gives

$$\hat{C}_n^2(h_j) = \sum_i W_{ji}^{-1}(C_i+N_i) \tag{14}$$

Thus the error variance in the mean $C_n^2(h_j)$ value is

$$\left\langle\left(\hat{C}_n^2(h_j) - C_n^2(h_j)\right)^2\right\rangle = \sum_i (W_{ji}^{-1})^2 \langle N^2 \rangle \qquad (15)$$

where we assume that $N_i N_j = \delta_{ij} N^2$. Since the correlations C_i will be taken sequentially, we assume that N_i and N_j for $j \neq i$ are uncorrelated. In practice each measurement will take roughly one minute, thus this assumption should be quite good.

The noise analysis of this measurement instrument now reduces to doing the integrals numerically to find W_{ij}, inverting to get W_{ji}^{-1}, summing the squares of W_{ji}^{-1}, and multiplying by N^2. A computer program has been written to calculate $\sum_i (W_{ji}^{-1})^2$ and the results are presented in figure 2.

N^2 Estimate

An approximate expression for $\langle N^2 \rangle$ can be found by dividing the variance of an instantaneous measurement by the number M of independent measurements during the duration T of the correlation interval. The variance of an instantaneous measurement is

$$\sigma_{INST}^2 \simeq \left\langle\left[\frac{\partial\phi_1}{\partial x} \; \frac{\partial\phi_2}{\partial x}\right]^2\right\rangle - \left\langle\frac{\partial\phi_1}{\partial x} \; \frac{\partial\phi_2}{\partial x}\right\rangle^2 \qquad (16)$$

Now $\frac{\partial\phi_1}{\partial x}$ and $\frac{\partial\phi_2}{\partial x}$ are somewhat correlated. Thus we calculate σ_{INST}^2 for the two extremes of perfect correlation and independence. For independence, we get

$$\sigma_{INST}^2 = \left\langle\left[\frac{\partial\phi_1}{\partial x}\right]^2\right\rangle^2 \qquad (17)$$

$$\simeq \left\langle\frac{[\phi(L) - \phi(0)]^2}{L^2}\right\rangle^2$$

$$= D^2(L)/L^4$$

where L is the aperture diameter and where D(L) is the structure function of the turbulence defined by

$$D(L) = \langle |\phi(X+L) - \phi(X)|^2 \rangle \qquad (18)$$

$$= 2.91 \int_0^\infty dh \; C_n^2(h) \; L^{5/3} \sec\Theta_o$$

where the $\sec\Theta_o$ corrects for the longer slant path through the atmosphere.

In the case of perfect correlation we can calculate σ_{INST}^2 by assuming $\frac{\partial\phi_1}{\partial x}$ is a gaussian variable with variance $\left\langle\left[\frac{\partial\phi_1}{\partial x}\right]^2\right\rangle$. In this case

$$\sigma_{INST}^2 = 2\left\langle\left[\frac{\partial\phi_1}{\partial x}\right]^2\right\rangle^2 \simeq 2D^2(L)/L^4 \qquad (19)$$

Thus the dependence or independence of $\frac{\partial\phi_1}{\partial x}$ and $\frac{\partial\phi_2}{\partial x}$ has the effect of a factor of 2. For this analysis we assume the worst case in order to get a pessimistic noise figure.

Now to get N^2 we must divide σ_{INST}^2 by the number M of independent phase slopes in the measurement time T.

One can estimate M by saying that the slopes are reasonably independent when the turbulence has shifted by 2 aperture diameters. If the near velocity of the turbulence (due to slew and/or wind) is v, then

$$M = \frac{Tv}{2L} \qquad (20)$$

where L is the aperture size in the direction of the turbulence motion.

As an example, for v = 5 meters/second, L = .01 meters, T = 60 seconds, then M = 15000 independent measurements.

Example

In order to get a sense of how well this instrument operates, the numerical integrations have been performed for a variety of parameter values for an atmosphere corresponding to a coherence length r_o = 10 centimeters for .5 micron light at a zenith angle of $\Theta_o = 60°$. Using the formula for D(L) in terms of r_o, we get from eq.19

$$\sigma_{INST}^2 \simeq \frac{2 \times 6.88\lambda^2}{(2\pi)^2 r_o^{5/3} L^{1/3}} \sec\Theta_o \qquad (21)$$

where we used $D(L) = \frac{6.88}{(2\pi/\lambda)^2} (L/r_o)^{5/3}$ for a vertical line of sight.

Then the stastical error $\langle N^2 \rangle$ is just σ^2_{INST} divided by the number M of independent measurements in one correlation period. The tabulated computer runs in figure 2 list $[\sum_i (W_{ij}^{-1})^2]^{\frac{1}{2}}$ for λ = .4 microns and $\bar{\lambda}$ = .6 microns, and must be multiplied by $\langle N^2 \rangle^{\frac{1}{2}}$ in order to get the standard deviation of the measurement. All turbulence is assumed to lie in the range from 0 to 12000 meters altitude. The value in the table depends somewhat on the particular level j trying to be measured, but the variation is not large from level to level so we list only the average value for the levels.

To list one case concretely, take N=5 levels, L_x=1 centimeter, L_y=4 centimeters with a 4% bandwidth for the light. The table lists $[\sum (W_{ij}^{-1})^2]^{\frac{1}{2}} = 4.4\text{x}10^{-11}$. For $M=10^4$ independent measurements (which we estimated would occur in about 60 seconds of correlation), the measurement error is

$$\sigma_{C_n^2} = \frac{(4.4\text{x}10^{-11})\ \sigma^2_{INST}}{10^2} \tag{22}$$

$$= 1.91\text{x}10^{-18}\ \text{meters}^{-2/3}$$

Since C_n^2 values in the atmosphere typically range up to 10^{-15} meters$^{-2/3}$, this measurement error is quite acceptable.

Photon Noise

In all this calculation, the contribution of photon noise to the error has been neglected. This is because that error is small compared to the statistical sampling error for a 1st magnitude star. If the correlation time of the atmosphere is 1 millisecond, then the number of photons received from a 1st magnitude star for one tilt measurement is about 40 (with a quantum efficiency of .1, an optics transmission of .5, and a 10% bandwidth around .5 microns).

The variance of the tilt measurement is

$$\left\langle \left| \frac{\partial \phi_1}{\partial x} \right|^2 \right\rangle \simeq \frac{\lambda^2 D(L)}{L^2} \tag{23}$$

$$\simeq 1\text{x}10^{-11}\ \text{radians}^2$$

for r_o = 10 centimeters, λ = .5 microns, L = 1 centimeter.

The photon noise on this measurement for a Hartmann sensor using a quad cell to detect the shift of the tilted wavefront is

$$\sigma^2_{ph} = \frac{\lambda^2}{4\pi^2 N L^2} \qquad (N = \#\text{ of photons}) \tag{24}$$

$$\simeq 1\text{x}10^{-12}\ \text{radians}^2$$

which is about ten times less than the statistical error variance. If the atmospheric correlation period is 4 milliseconds, then the photon error is 40 times less than the statistical error. Since the system has great freedom in choosing its sources, a suitably bright source can be found which will give performance limited only by atmospheric statistics.

Conclusions

The conclusion from this analysis is that this new instrument using atmospheric dispersion is capable of resolving 5 or more layers of turbulence with fair speed and accuracy. The bandwidth of the light used in the measurement is important (see figure 2) and must be narrower to resolve more layers. Similar conclusions apply to the subaperture dimensions L_x and L_y. Photon noise from a 1st magnitude star is not seen to be a major limitation since statistical sampling error is significantly larger. A typical case gave an accuracy of $2\text{x}10^{-18}$ meters$^{-2/3}$ for measuring C_n^2 resolved into 5 layers of atmosphere in 60 seconds correlation time.

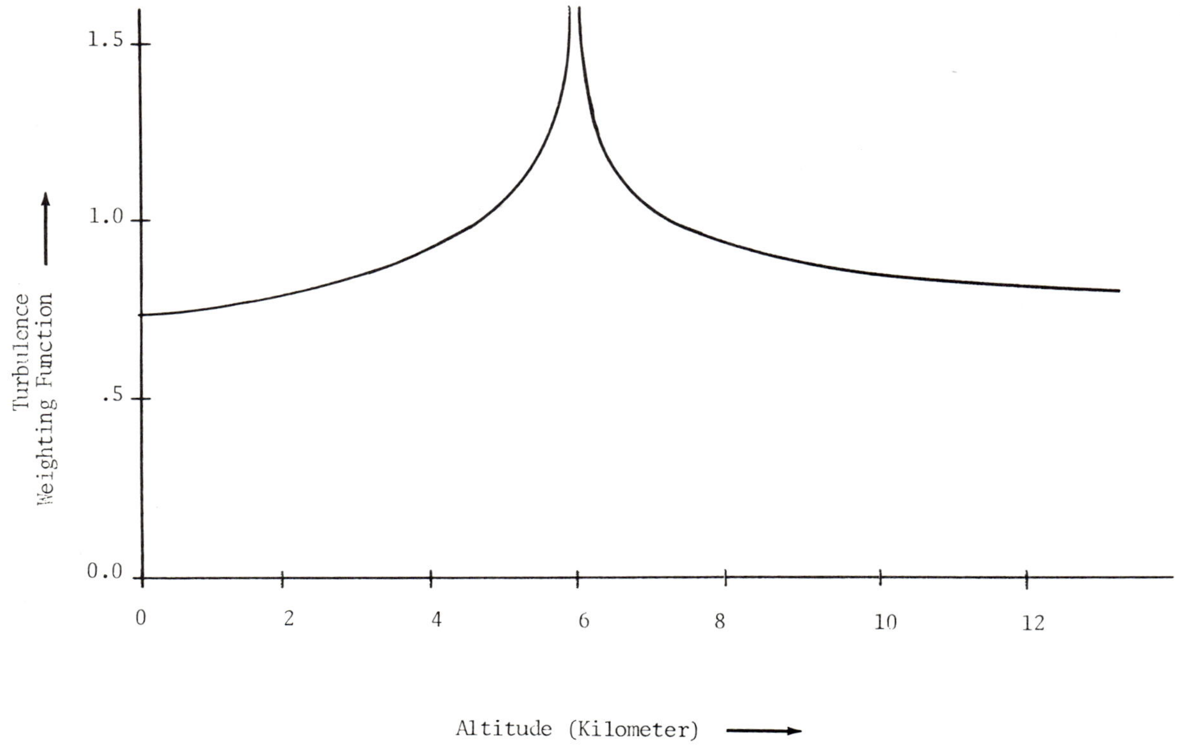

Figure 1: Plot of the Turbulence Weighting Function for 400 nanometer and 600 nanometer Wavelengths Using a Displacement of 6 centimeters in the Subapertures.

Figure 2

Parameter Tradeoffs

	N Levels	L_x	L_y	Fractional Bandwidth	$[\sum_i [W_{ji}^{-1}]^2]^{1/2}$ Error Coefficient
L_y Tradeoff	5	.01	.01	.1	1.0×10^{-10}
			.02		1.3×10^{-10}
			.04		1.8×10^{-10}
			.06		2.3×10^{-10}
L_x Tradeoff	5	.001	.05	.1	1.4×10^{-10}
		.01			2.0×10^{-10}
		.02			10.3×10^{-10}
Bandwidth Tradeoff	5	.01	.04	0	$.36 \times 10^{-10}$
					$.36 \times 10^{-10}$
					$.36 \times 10^{-10}$
					$.36 \times 10^{-10}$
					$.44 \times 10^{-10}$
					$.94 \times 10^{-10}$
					1.8×10^{-10}
# Levels Tradeoff	3	.01	.04	.10	$.28 \times 10^{-10}$
	4				$.64 \times 10^{-10}$
	5				1.8×10^{-10}
	6				8.6×10^{-10}
	7				$34. \times 10^{-10}$
	8				520×10^{-10}

References

1. American Institute of Physics Handbook, McGraw Hill, New York, 1972, p. 3-7θ.

2. Ibid, p. 6-111.

3. Tatarski, V.I., Wave Propogation in a Turbulent Medium, Dover Publications, New York, 1961, Ch. 6.

TURBULENCE EFFECTS UPON LASER PROPAGATION IN THE MARINE BOUNDARY LAYER*

Kenneth L. Davidson and Thomas M. Houlihan
Naval Postgraduate School
Monterey, California 93940

Abstract

Shipboard measurements of small scale temperature and velocity fluctuations have been accomplished to determine optical wave propagation properties of the marine boundary layer. Measurements were recorded for ocean conditions in Monterey Bay and in the confines of the Pacific Missile Range. Turbulence parameters measured were the temperature structure function parameter, C_T^2, and rate of dissipation of turbulent kinetic energy, ε. There is satisfactory agreement with overland predictions for the variation of these parameters with height.

Introduction

On the basis of the isotropic nature of small scale fluctuations, only one parameter is necessary to describe the intensity of the atmospheric refractive index fluctuations over many scales.(1) It is the refractive index structure function parameter, C_N^2, where

$$C_N^2 = \overline{[n'(x) - n'(x+r)]^2}/r^{2/3} \tag{1}$$

Herein, n'(x) and n'(x+r) are refractive index fluctuations at two points on a line oriented normal to the mean wind direction separated by the distance, r. This distance, r, is less than the outer scale, L_0 - the lower end of the inertial subrange - and greater than the inner scale, ℓ_0 - the smallest scale of turbulence occurring naturally. The brackets in Eq. (1) designate an RMS evaluation of the quantities contained therein.

Fortunately, C_N^2 can be related to the temperature structure function parameter, C_T^2 in the following manner, i.e.,

$$C_N^2 = [79 \times 10^{-6}(P/T^2)]^2 \, C_T^2 \tag{2}$$

where P is the barometric pressure and T is the ambient temperature.

The innerscale (ℓ_0) is itself a relevant optical parameter of the atmosphere since it appears in the following expression for the Exp(-2) folding distance of the mutual coherence (ρ_0), a defining parameter in an expression for the mutual coherence function (2),

$$\rho_0 = 1/(1.478 \, C_N \, \ell_0^{-1/6} \, \bar{k}) \tag{3}$$

where C_N is the refractive index structure function (defined previously) and $\bar{k}$ is average wave number of the illumination.

Fortunately, ℓ_0 is also related to a measurable property of the atmospheric turbulence, it being the rate of dissipation of turbulent kinetic energy (ε), i.e.,

$$\ell_0 = (\gamma^3/\varepsilon)^{1/4} \tag{4}$$

where γ is the kinematic molecular viscosity.

Descriptions of the small scale meteorological turbulent fluctuation properties, such as C_T^2 and ℓ_0, which characterize laser propagation, have not been as complete nor in the quantity for the overwater regime as for the overland regime. Overwater descriptions are necessary, even though considerable progress has been made in overland investigations, such as from experiments by AFCRL described by Wyngaard, et al. (3) The necessity exists because of increasing evidence of influence on the airflow by the surface waves. The wave influence has been observed to be significant enough to warrant an evaluation of existing empirical expressions relating small scale properties to mean wind, temperature, and humidity profiles.(4) Observational experiments to evaluate existing predictions with regard to the overwater regime should emphasize the height dependence of small scale properties for different stability conditions.

*The research work leading to this paper was supported by the Naval Sea Systems Command (PMS-405).

For this evaluation, the small scale properties measured should include the temperature structure function parameter (C_T^2) and the inner scale (ℓ_0) of the velocity fluctuations. The inner scale, as indicated above, is related to the viscous dissipation rate of turbulent kinetic energy (ε). Since it is also necessary to describe mean thermal stratification in terms of the atmospheric stability parameters, such as Ri (Richardson Number) or L (Monin-Obukhov Length), which are defined in (3), measurements are also required of mean profiles of wind speed, temperature and humidity.

Instrumentation and Shipboard Mounting Arrangement

Shipboard observational experiments to describe the height and stability dependence of small scale turbulence properties, including C_T^2, were performed in Monterey Bay and over the open ocean off the west coast. These were completed in conjunction with simultaneous laser propagation experiments. The shipboard sensor mounting arrangement appears in Fig. 1 which shows the multi-level approach to evaluate height predictions. Instrumentation consisted of hot-wire anemometers, platinum resistance wire thermometers, and lyman-alpha humidiometers for fluctuating properties, and cup anemometers, quartz thermometers, and lithium chloride sensors for mean profile properties. Coincident acceleration measurements in early experiments supported other results that indicate the high frequency portion of velocity variance spectra can be interpreted for dissipation rates and inner scale, since ship related motion affects the variance spectra at frequencies below those of interest.

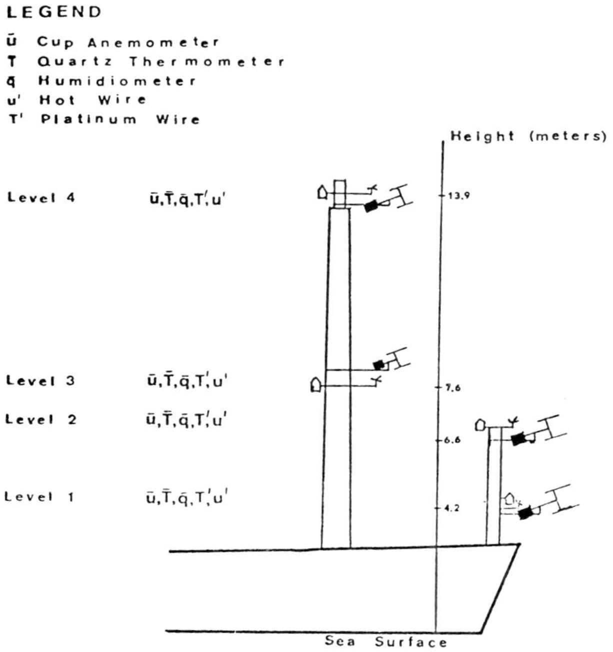

Fig. 1. Sensor mounting arrangement aboard R/V ACANIA (operated by Department of Oceanography, NPS)

The four Thornthwaite cup anemometer units featured lightweight plastic sensors to insure proper response at low windspeeds. The pulse outputs of the anemometer units were counted for ten minute periods and final counts were displayed for recording purposes using Hewlett-Packard 5221A Counters.

Temperature and humidity data were recorded by multiplexing sensor signals from each level through respective readout units and recording the data in digital format upon an HP 561B Printer.

The temperature transducers utilized were HP 2833 Oceanographic (quartz crystal) Sensors set for resolution of 0.001°C. The sensors and their accompanying booster oscillators were sequentially multiplexed through an HP 2801 Quartz Thermometer Readout Unit for monitoring.

The humidity sensors utilized were Hygrodynamics, Inc. Model 1818W wide range sensors which featured an integral thermistor probe having a resolution of 0.1°C. These units were sequentially multiplexed through a Hygrodynamics Digital II Readout Unit for monitoring of both relative humidity and temperature data. Both quartz temperature sensors and humidity sensors at each level were contained within Hygrodynamics 15-6175 Weather Shelters to provide for proper operation in the marine environment.

Fluctuating velocity signals were recorded using Thermosystems 1210W-T1.5 probes and Thermosystems Model 1054B anemometer units. Fluctuating temperature signals were recorded using Thermosystems 1210W-P.8 probes and Sylvania Model 140 Thermosonde units which were adapted for shipboard use. Fluctuating humidity signals were recording using ERC Model BR Lyman Alpha Humidiometers.

All fluctuating signals were recorded on a 14-channel Sanborn Model 3950 Intermediate Band Analog Tape Recorder for later processing. Spectrum analyses were accomplished with a Nicolet Scientific UA-500 Analyzer with final scaled output prepared on an HP 7035 X-Y Recorder.

Results

C_T^2 results.

Examinations of the temperature structure function parameter, C_T^2 , were made from both structure function measurements, using separated pairs of sensors. The expressions relating C_T^2 to the structure function, $D_T(r)$, and temperature variance spectrum, $\phi_T(k)$, are

$$D_T(r) = \overline{[T(x) - T(x+r)]^2} = C_T^2\, r^{2/3} \qquad (5)$$

where r is the sensor separation distance and T is temperature and

$$\phi_T(k) = \beta C_T^2\, k^{-5/3} \qquad (6)$$

where β is an empirical constant and k is the wave number.

An analysis of similarity predictions by Wyngaard et al. yielded the following general relations between C_T^2 and the stability parameters L or Ri

$$C_T^2 = T_*^2\, Z^{-2/3}\, f_2(Z/L) \qquad (7)$$

$$C_T^2 = Z^{4/3}\left(\frac{\partial\bar{\theta}}{\partial Z}\right)^2 f_1(Ri) \qquad (8)$$

where Z is the height above the surface, $(\partial\bar{\theta}/\partial Z)$ is the mean temperature gradient, T_* is the scaling temperature defined by the sensible heat and momentum transfer and which is independent of height and f_1 and f_2 are empirical functions which were determined experimentally by Wyngaard et al. Predictions from their empirical functions will be used in presenting overwater data in the following discussion.

First, C_T^2 results were obtained from structure function measurements at two levels above mean water level, 26 ft. and 38 ft. These yielded a consistent height dependence for C_T^2 of $Z^{-4/3}$ over a long period during which C_T^2 at individual levels changed by a factor of 3 between minimum and maximum values. These results are shown in Fig. 2 in the form of the ratio $C_T(38)/C_T(26)$ for which a value of .776 would correspond to C_T^2 having a $Z^{-4/3}$ height dependence. Although the $Z^{-4/3}$ height dependence was predicted by Wyngaard et al. for the free convection regime, these overwater results were observed in only slightly unstable conditions.

Several observational periods with near neutral thermal stratification yielded small signals for the temperature differences between the sensors in the pair. This led to the use of single sensors and temperature variance spectra (Eq. 6) for determining C_T^2. A composite of results from these measurements and analyses for up to four levels appears in Fig. 3 in a log Z versus Log C_T^2 format. The composite shows an approximate $Z^{-4/3}$ slope in this representation for unstable conditions. A different slope, close to a $Z^{-2/3}$ dependence, describes the composite for stable conditions. Results of Wyngaard, et al., predicted a $Z^{-2/3}$ height dependence for neutral conditions and less than $Z^{-2/3}$ for stable conditions.

A comparison of the observed results and the Wyngaard et al. prediction (curve) based on Eq. (8) appears in Fig. 4. In Fig. 4, individual data points appear as dots and averages over Richardson number (Ri) intervals of .25 appear as dots within a larger circle. The error bars are standard deviations from the mean within each interval. The number at the top of error bars is the observations defining the mean value. The average values agree satisfactorily with the predicted curve for negative Richardson numbers; the unstable stratification case when the air density decreases with height. There is a considerable difference between the predicted curve and the average values for positive Richardson numbers; the stable stratification case. It is noted that if the Richardson number exceeds a critical value, turbulence is expected to be virtually non-existent due to the extreme stability. This critical value generally is considered to be .21.

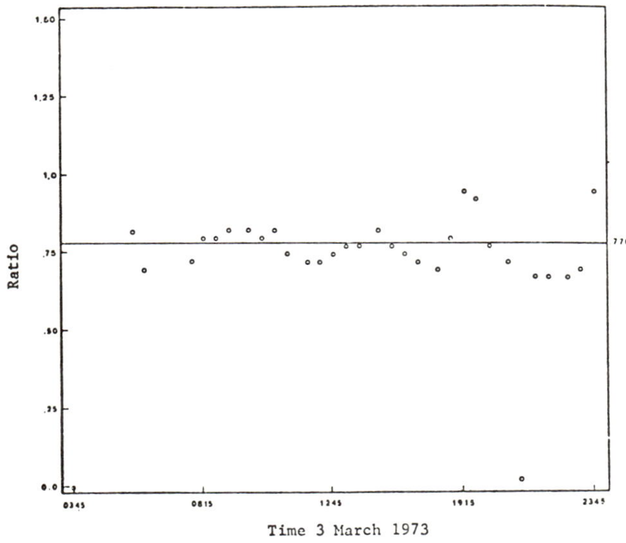

Fig. 2 Thirty-minute averages of $C_T(38ft)/C_T(26ft)$ versus time (PST)

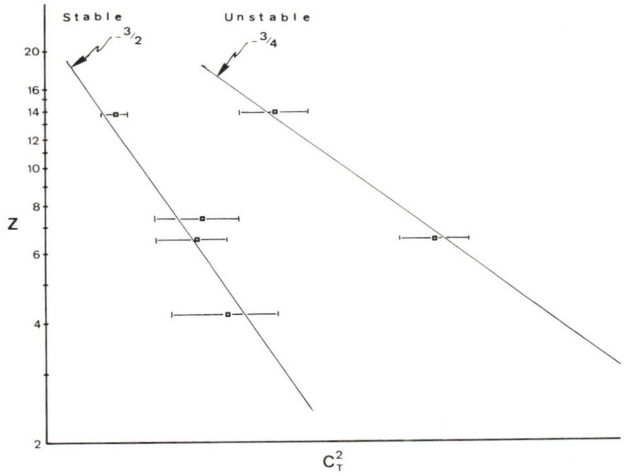

Fig. 3 Composite plots of C_T^2 versus height in meters, Z , includes nine periods.

The scatter in the observed results in Fig. 4 can be attributed to scatter in both the measured C_T^2 values as well as $\partial\theta/\partial Z$ values. The latter is important because we are dealing with small temperature gradients, occasionally, and the quantity is squared for the normalization.

ε Results

Small scale velocity fluctuations are used in characterizing the overwater regime with respect to propagation effects through the innerscale, ℓ_0 , which is related to the viscous dissipation rate of turbulent kinetic energy, ε , by Eq. 4.

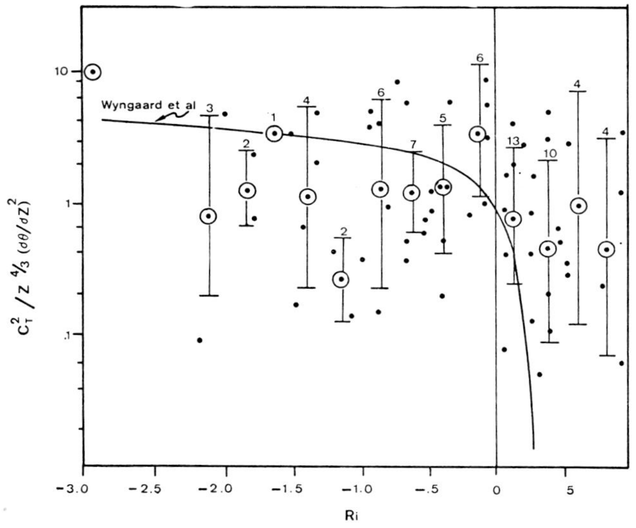

Fig. 4 Normalized C_T^2 values versus Richardson number, Ri , and comparison with prediction in (3).

The viscous dissipation rate, ε , was determined from variance spectral estimates, $\phi_u(k)$, in the inertial subrange using Kolomogorov's relation

$$\phi_u(k) = \alpha\varepsilon^{-2/3} k^{-5/3} \tag{9}$$

where α is an empirical constant and k is the wave number.

The velocity variance spectra exhibited distinct inertial subrange regions, identified by -5/3 slopes on log $\phi_u(k)$ vs. log k plots. Slope changes were evident at higher wave numbers corresponding to the lower limit of the dissipation subrange. The viscous dissipation rate, which also was examined for height variations, was computed from spectral estimates in the inertial subrange, using Eq. 9 and also from inner scale estimates, determined from wave numbers corresponding to the lower limit of the dissipation subrange. A significant result of this was that ε values, estimated by the two methods, were in good agreement. An example of this agreement is shown for one period in Fig. 5 in a log ε versus log Z format.

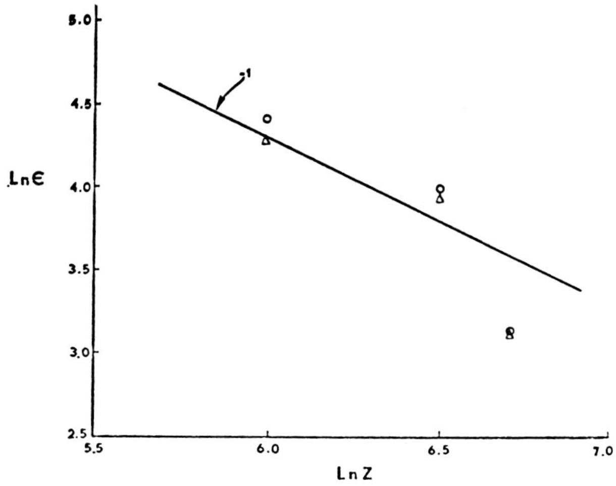

Fig. 5 Comparison of ε estimates; inertial subrange O, dissipation subrange Δ.

Analysis of similarity predictions for the turbulent kinetic energy balance for a steady, horizontally homogeneous and non-divergent boundary layer yields the following expression for ε[5]

$$\varepsilon = k u_* Z^{-1} f_3(Z/L) \quad (10)$$

In Eq. 10, k is the von Karman constant, u_* is the friction belocity which is independent of height, (u_*^2 is the momentum transfer per unit mass), and f_3 is an empirical function previously determined in several investigations, most recently by Businger et al.[6]

Composites of ε versus height were examined for evidence of expected stability influence. For near neutral conditions when the predicted value of f_3 is equal to 1, the height dependence is expected to be Z^{-1} based on the following relation

$$\varepsilon = u_*^3 (kZ)^{-1} \quad (11)$$

Existing empirical relations[6] for f_3 in Eq. 10 predict that the absolute value of the exponent on Z should be larger than 1 in unstable conditions and that it should be less than 1 in stable conditions. Composites for unstable and stable conditions appear in Fig. 6 in a log Z versus log ε format. As seen from Fig. 6, a stability influence on the height variation of ε cannot be discerned in the results. However, they do vary around the Z^{-1} prediction for neutral conditions.

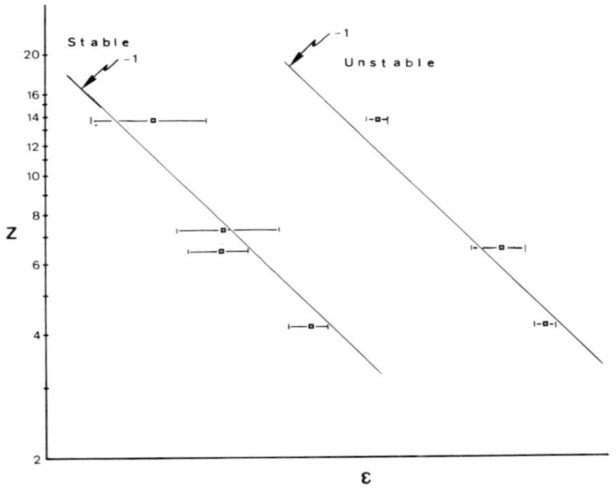

Fig. 6 Composite plots of ε versus height in meters, Z , includes nine periods.

Conclusions

A shipboard system is providing results which can be interpreted for descriptions of small scale turbulence properties in the marine boundary layer, properties which are relevant in analyzing optical wave propagation and image resolution. The existing results appear to agree partially with predictions based on an overland turbulence and profile investigation by Wyngaard et al.[3] However, the scatter occurring when comparisons are made with the predicted curve over a wide Richardson number range indicates that the waves' presence may have a significant influence. Additional data is necessary to describe differences which occur over water due to the wave influence.

References

1. Tatarski, V. I., "Wave Propagation in a Turbulent Medium," (McGraw-Hill, New York, N. Y.), Ch. 10, 1961.

2. Livingston, P. M., "Study of Target Edge Response Viewed Through Atmospheric Turbulence Over Water," Applied Optics 11:10, 1972.

3. Wyngaard, J. C., Izumi, Y., and Collins, S. A. Jr., "Behavior of the Refractive Index-Structure Parameter Near the Ground," J. Opt. Soc. Am. 61:12, 1971.

4. Davidson, K. L., "Observational Results on the Influence of Stability and Wind-Wave Coupling on Momentum Transfer and Turbulent Fluctuations Over Ocean Waves," Boundary-Layer Meteorology 6, 1974.

5. Garratt, J. R., "Studies of Turbulence in the Surface Layer Over Water (Lough Neagh). Part II. Production and Dissipation of Velocity and Temperature Fluctuations," Quart. J. R. Met. Soc. 98:417, 1972

6. Businger, J. A., Wyngaard, J. C., Izumi, Y. and Bradley, E. F., "Flux-Profile Relationships in the Atmospheric Surface Layer," J. Atmospheric Sci. 28:2, 1971.

Session 3
SPECKLE INTERFEROMETRY

Session Chairman
R. Wagner
Optical Sciences Center, University of Arizona

HOW TO BUILD A SPECKLE INTERFEROMETER*

Arthur M. Schneiderman and Douglas P. Karo
Avco Everett Research Laboratory, Inc.
Everett, Massachusetts 02149

Abstract

This paper briefly reviews the current status of speckle interferometry including the recent extension proposed by Knox and Thompson and the limitations imposed by non-isoplanicity. The speckle interferogram is characterized in terms of its scale lengths and photon statistics. The various subsystems of the instrument are reviewed in detail in terms of their required performance. The overall S/N is defined and discussed in terms of both unresolved and extended objects.

Introduction

Since the introduction of Speckle Interferometry (SI) in 1970[1] as a means for overcoming atmospherically induced aberrations in large aperture imaging systems, SI has received a great deal of attention from the astronomy and optics communities. However, aside from the work of its originator, A. Labeyrie, and his co-workers, there has been little significant published data obtained by this technique.[2-5] This is no doubt related to the fact that there has also been no systematic description in the literature of the Speckle Interferometer itself. In this paper we shall attempt to present such a description.

To those persons versed in image processing, SI is directly described by the simple equation

$$\langle |\tilde{I}(\vec{f})|^2 \rangle = |O(\vec{f})|^2 \langle |\tilde{\tau}(\vec{f})|^2 \rangle \qquad (1)$$

where $\tilde{I}(\vec{f})$ and $O(\vec{f})$ are the Fourier transforms of the degraded and perfect images $\tilde{I}(\vec{x})$ and $O(\vec{x})$, respectively, while $\tau(\vec{f})$ is the combined lens-atmosphere optical transfer function. $|(\ )|$ denotes the modulii of the quantities while $\langle(\ )\rangle$ signifies their ensemble average. The ensemble average (a statistical concept) is required since $\tilde{\tau}(\vec{f})$ is random, as $\tilde{(\ )}$ denotes.

Inherent in Eq. (1) are a number of important assumptions about the lens-atmosphere combination as a transfer element between the object and the image. One important aspect, which is receiving increased interest is the required spatial invariance of $\tilde{\tau}(\vec{f})$, commonly referred to as isoplanicity.[6]

The SI process involves the following steps:

(1) Obtain suitable short exposure images.

(2) Fourier transform these images and determine the modulus of this complex function.

(3) Co-add a number of these modulii.

(4) Obtain the lens-atmosphere modulation transfer function $\langle |\tilde{\tau}(\vec{f})|^2 \rangle$. This can be done by repeating (1)-(3) using an unresolvable (point) object, making certain that spatial (isoplanicity) <u>and</u> temporal stationarity are satisfied for this quantity.

(5) Invert Eq. (1) to determine $|O(\vec{f})|^2$ or its Fourier transform $R(\vec{\xi})$, the object's autocorrelation function.

(6) Interpret $|O(\vec{f})|^2$ or $R(\vec{\xi})$ to obtain relevant object information.

This paper will deal mainly with item (1). We will determine what a "suitable" image is and identify at least one means of achieving it.

The second step, Fourier transforming, can be done optically,[1,7] digitally,[8] by a number of analog methods,[9] or by combinations of these. Each method has different advantages with respect to speed (real time capability), accuracy, precision, flexibility, linearity, etc. There is an extensive and easily accessible literature on this subject which has been developed over the past twenty-five years. We shall, therefore, assume that the basis for this step can be determined elsewhere.

The next three steps are self-explanatory while the final step represents the most challenging aspect of current research on SI as a technique.

---

*This work was supported in part by the Defense Advanced Research Projects Agency under Contract F04701-75-C-0047.

In addition to SI, two related techniques have been successfully employed in overcoming atmospheric aberrations: Amplitude Interferometry (introduced by Michelson) and Intensity Interferometry (Hanbury-Brown and Twiss). All three techniques have as their final "output" $|O(\vec{f})|$. It is generally believed that knowledge of $|O(\vec{f})|$ alone is insufficient for image reconstruction, i.e., since $O(\vec{f}) = |O(\vec{f})| \, e^{i\,\Phi_o(\vec{f})}$, one needs to also determine $\Phi_o(\vec{f})$ in order to generate an image by inverse transformation. Some theoretical efforts suggest that $\Phi_o(\vec{f})$ can be deduced from $|O(\vec{f})|$ through the use of dispersion relations although this remains undemonstrated.

Both theory[10] and experiment show that above the seeing limited spatial frequency $f_s = r_o/\lambda f$, (where r_o is the correlation scale of the atmospherically induced aberrations at the aperture, λ is the wavelength, and f is the effective focal length of the telescope), $\langle \tilde{\Phi}(\vec{f}) \rangle \cong 0$ where $\tilde{\Phi}(\vec{f}) = \Phi_o(\vec{f}) + \tilde{\Phi}_A(\vec{f})$, $\tilde{\Phi}_A(\vec{f})$ being the atmospheric (random) contribution to $\tilde{\Phi}(\vec{f})$. Therefore, direct processing to determine $\Phi_o(\vec{f})$ does not work.

Recently, Knox and Thompson[11] (K-T) have proposed that the quantity $\Delta\Phi(\vec{f}) = \langle \Phi(\vec{f}) - \Phi(\vec{f}+\vec{\Delta}) \rangle \cong \vec{\Delta}\Phi_o(\vec{f})$ can be determined in the presence of the atmosphere and used to find $\Phi_o(\vec{f})$ and then combined with $|O(\vec{f})|$ to generate an image. This subject is currently under active study by a number of researchers.

From what is currently known about K-T, the criteria for producing useful data for this algorithm are identical with those for SI. We shall therefore, for the present purposes, consider K-T as an extension of SI.

For a wide class of simple objects, knowledge of $|O(\vec{f})|$ fully specifies the object. This class consists of all center-symmetric objects (i.e. $O(\vec{x}) = O(-\vec{x})$). Included are equal magnitude binary stars, unresolvable stars, resolvable stars, etc. In practice, systems where a left-right ambiguity is tolerable are also common (unequal magnitude binary stars). It is these objects which have given the first useful demonstrations of SI.[2-5]

Application of SI to more complex shapes requires the invention of decoding techniques to obtain unambiguous object information from knowledge of the autocorrelation function. One exception would be the fortuitous combination of a bright point object sufficiently separated from the object of interest, but still within the relevant isoplanatic patch. In this unusual case, portions of the total autocorrelation function would contain images of the object of interest.[12]

As part of step (6), one must also consider the effect of failure (to some extent) of Eq. (1). For example, spatial non-isoplanicity and spatially varying object brightness produce similar effects on $|O(\vec{f})|$. For binary stars, the fringe modulation of $|O(\vec{f})|^2$ is given by

$$\frac{|O(\vec{f})|^2_{max} - |O(\vec{f})|^2_{min}}{|O(\vec{f})|^2_{max} + |O(\vec{f})|^2_{min}} = \frac{2m_1m_2}{m_1^2+m_2^2} \; \frac{\langle \tilde{\tau}_1(\vec{f})\, \tilde{\tau}_2^*(\vec{f}) \rangle}{\langle |\tilde{\tau}(\vec{f})|^2 \rangle} \; . \qquad (2)$$

m_1 and m_2 are the brightness, τ_1 and τ_2 the instantaneous transfer functions for each component, and $\langle|\tau_1|^2\rangle = \langle|\tau_2|^2\rangle \equiv \langle|\tau|^2\rangle$. $\langle \tau_1\tau_2^* \rangle / \langle|\tau|^2\rangle = M(\vec{f})$ is a measure of non-isoplanicity. Korff[6] has shown that above the seeing limit, f_s, $M(\vec{f})$ is nearly independent of $\vec{f}$. Therefore, in this particular case, non-isoplanicity and different component magnitudes produce indistinguishable effects. In carrying out step (6), one must beware of these types of extraneous effects.

Non-isoplanicity is currently considered a limitation on object size. However, it is likely that a modified version of Eq. (1) can deal with finite amounts of non-isoplanicity. There has been substantial work in the statistics community on such non-stationary random processes. In fact, the original use of the structure function in propagation theory grew out of early work on non-stationary processes with stationary first increments. It is worth noting that there are, as of this date, no published measurements of atmospheric non-isoplanicity.

With this general background, let us now return to step (1) and characterize the "suitable" speckle interferometer.

Characteristics of Speckle Interferograms

Before developing the design criteria for a speckle interferometer, let us first characterize the image we are trying to detect.

Figure 1 contains a realistic simulated image[7] of an unresolved and an extended object. In designing a speckle interferometer it is worthwhile to keep these images in mind.

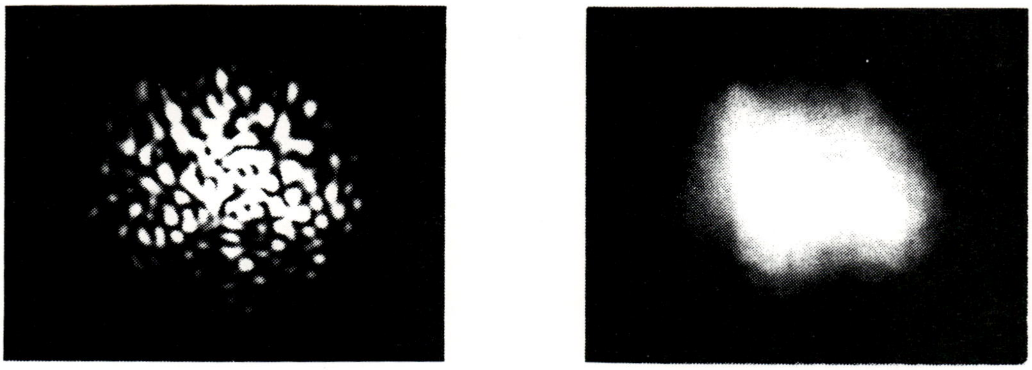

POINT OBJECT (a)

EXTENDED OBJECT (b)

Fig. 1. Typical Speckle Interferograms.

We can characterize these images, in an engineering sense, based on detailed theory, field data, and laboratory simulation. For this purpose we describe the speckle image of a point object by

$$\tilde{I}_p(\vec{x}) = \langle \tilde{I}_p(\vec{x}) \rangle \,(1 + \tilde{\chi}(\vec{x})) \tag{3}$$

where $\tilde{\chi}(\vec{x})$ is an exponentially distributed random variable of zero mean and unit variance and $\langle \tilde{I}_p(\vec{x}) \rangle$ is the mean image corrected for image wander (i. e. Fried's "short exposure" image[11]).

The characteristic size of $\langle \tilde{I}_p(\vec{x}) \rangle$ is $\lambda f/r_o$ where, again, r_o is the conventional atmospheric correlation scale, and is nominally 10 cm corresponding to $1''$ seeing. The correlation scale of $\tilde{\chi}(\vec{x})$ is $\lambda f/D$, the telescope's diffraction limited spot size, where D is the telescope diameter.

A short exposure point object image can, therefore, be described as having a size $\lambda f/r_o$, a finest scale $\lambda f/D$ and a contrast

$$C_p \equiv \frac{\langle (\tilde{I}_p(\vec{x}) - \langle \tilde{I}_p(\vec{x}) \rangle)^2 \rangle}{\langle I_p(\vec{x}) \rangle^2} \tag{4}$$

of 100%.

An extended object of angular size $\underline{S}$, will produce an image which is the convolution of the point object image with a perfect geometric image of the object. The resulting image contrast will be reduced to roughly

$$C_{\underline{S}} \cong \frac{1}{1 + \left(\frac{\underline{S}}{\lambda/D}\right)^2} \tag{5}$$

Next, let us determine the image irradiance. Since most small extra terrestrial objects are very dim, we will formulate this concept in terms of photon fluxes. It is useful both for instrument design and for later analysis of signal to noise ratio to determine the average expected number of "photons per speckle", n

An unresolved star of apparent photometric magnitude m will produce, at the aperture of a telescope, an average irradiance

$$I_p \cong 3 \times 10^{-8} \ \Delta\lambda \ e^{-0.92m} \ \text{watts/m}^2 \tag{6}$$

where $\Delta\lambda$ is the bandwidth in microns. Thus, assuming no imaging losses and a clear aperture, the total image contains

$$N_p = I_p \frac{\pi D^2}{4} \frac{\lambda}{hc} t_{exp} \quad \text{photons} \tag{7}$$

where hc/λ is the energy per photon and t_{exp} is the exposure time (as will be discussed later, a short exposure time must be chosen to effectively freeze the changing speckle pattern and allow the recording of images with diffraction limited spatial frequency content). Since an image contains of order $(D/r_o)^2$ speckles, we find, approximately,

$$n_p \cong I_p \frac{\pi r_o^2}{4} \frac{\lambda}{hc} t_{exp} \tag{8}$$

Note that n_p is independent of D, so that, independent of its size, any large telescope will produce the same number of photons per speckle. For $\lambda = 0.5\mu$, $\Delta\lambda = 0.025\mu$, $t_{exp} = 10^{-2}$ sec, $r_o = 10$ cm and $m = 5$, this relation yields $n_p = 1500$ photons/speckle. For poorer seeing ($r_o < 10$ cm) or faster seeing ($t_{exp} < 10^{-2}$ sec) this quantity is reduced according to the above equation.

For an extended object, we note that all points within a projected seeing spot at the object will contribute at a point in the image plane. If we now define m in Eq. (6) as the average apparent magnitude per diffraction limited resolution element at the object, then the number of photons per speckle increases with increasing $\underline{S}$ as $(1 + (\underline{S}/\lambda/D)^2)$ to a limiting value of $1 + (D/r_o)^2$ when $\underline{S} \geq \lambda/r_o$.

This, then, completes our description of the size, structure and brightness of typical speckle images.

Requirements of a Speckle Interferometer

A speckle interferometer is, basically, a high quality, diffraction limited camera which must be capable, in itself, of producing near diffraction limited images in the absence of the atmosphere. The presence of the atmosphere places added requirements for dispersion correction, shuttering (at high duty cycle) and wavelength filtering. The statistical nature of SI and the form of the data analysis relaxes some of the classical requirements in terms of sensor linearity, sensitivity uniformity, and dynamic range.

The functional components of a speckle interferometer are:

1. a magnification system to match the telescope plate scale with the sensor resolution,
2. a shutter to permit "short" exposures,
3. a system for correction of atmospheric dispersion,
4. a wavelength filtering system (λ_o and $\Delta\lambda$) to maintain coherence and reduce diffraction blurring,
5. a high quality, low light level (L^3) sensor,
6. a matched recording system (or real time data processor).

The Magnification System

Based on the previous section, a telescope of focal length will produce an image of size $\lambda f/r_o$ for objects unresolved on the seeing limited scale or $\underline{S} f$ for extended objects. In either case, the highest spatial frequency present in the image will correspond to the diffraction limit and is $f_{D.L.} = D/\lambda f$. For a typical astronomical telescope of F/16 (i.e., $f/D = 16$) this corresponds to $f_{D.L.} \cong 125$ cycles/mm.

Since the sensor should oversample by at least a factor of two, (four being a more common choice), one requires a sensor with a reasonably flat MTF to at least 250 cycles/mm. Few sensors combine this required high resolution with adequate sensitivity. Therefore, one generally chooses the sensor first and then uses a magnification lens to reduce $f_{D.L.}$ to more appropriate values.

A competing consideration is field of view (FOV). As the magnification m is increased, $f_{D.L.}$

decreases proportionally as does the FOV. In general, however, we are interested in viewing objects, or pieces of objects (by suitable masking) that are close to the seeing limit in size, i. e., the normally required FOV is a few times λ / r_o . For the 200 inch Hale telescope, this leads to a sensor requirement of approximately 500 x 500 pixels for a 5 FOV.

Both image intensified cine and low light level video cameras have been utilized in speckle interferometers. Typical MTF's for these systems provide adequate performance to ~20 cycles/mm.

In summary, we require a magnification of

$$M \simeq \alpha \frac{1}{\lambda F} \frac{1}{f_{c.o.}} \tag{9}$$

where $f_{c.o.}$ is the cut-off spatial frequency of the sensor (~20 cycles/mm) and α is the sampling rate (pixels per λF) usually taken as 2 to 4. The FOV of this system will be

$$\begin{aligned} FOV &= \frac{\text{diameter of sensor}}{f M} \\ &= (\text{diameter of sensor}) \frac{\lambda f_{c.o.}}{D} \end{aligned} \tag{10}$$

with a 100" diameter, F/16 telescope using a 1" diameter, 20 cycle/mm sensor, $\alpha = 2$ and $\lambda_o = .5\mu$, we require $M = 12.5$ and obtain a $FOV = 10''$.

The Shutter

A speckle interferogram is a short exposure image which must be obtained in a time during which motion of the turbulent atmosphere is essentially "frozen". Depending on a large number of variables (e. g., time of day, season, geographic location, micrometerological conditions, etc.) the atmospheric time constant, t_{atm}, is generally assumed to range from 1 to 100 msec. There is growing evidence that the atmospheric time may be less than 1 msec on some occasions.

In Figure 2, we show the effect of excessive exposure times on $\langle |\tilde{\tau}(\vec{f})|^2 \rangle$ based on laboratory simulation. This result is consistent with other published estimates.(14) As the ratio t_{exp}/t_{atm} increases above unity, the measured values of $\langle |\tilde{\tau}(\vec{f})|^2 \rangle_m$ gradually change from $\langle |\tilde{\tau}(\vec{f})|^2 \rangle$ to $|\tilde{\tau}(\vec{f})|^2$, the "Fried Short Exposure" (13) lens-atmosphere MTF. The transition occurs by a gradual and surprisingly uniform disappearance of $\langle |\tilde{\tau}(\vec{f})|^2 \rangle_m$ between the seeing and diffraction limits. This suggests that in some cases moderately long exposure times may be used which lead to depressed transfer functions but better overall S/N (see below).

It would therefore appear prudent to design into a speckle interferometer a shutter capable of exposures from 0.1 to 100 msec. In operation, t_{exp} would then be chosen by

a. separate, simultaneous measurement of t_{atm}
b. trained visual inspection of the shuttered images
c. on line computation of $\langle |\tau(\vec{f})|^2 \rangle$

Commercially available electro-mechanical shutters easily cover the range 1 to 100 msec. Shorter exposures require some sort of electronic gating of the sensor. (Our experience indicates that this can adversely affect the performance of the sensor--be sure to check this.) For short exposures, the shutter should be located near a focal plane to avoid unwanted aperture apodization.

Dispersion corrector

Except at zenith viewing, the atmosphere acts, on the average, as a dispersive prism, becoming stronger as the horizon is approached. The basic calculation of dispersion is straightforward. Starting from the refraction angle as a function of elevation angle, dispersion is introduced, corrected for the altitude of the observatory. Thus

$$\theta_{dispersion} = \theta_{refraction}(\text{elevation}) \times \frac{n(\lambda_o - \frac{\Delta\lambda}{2}) - n(\lambda_o + \frac{\Delta\lambda}{2})}{n(\lambda_o)} \times \frac{p(H)}{p(O)} \tag{11}$$

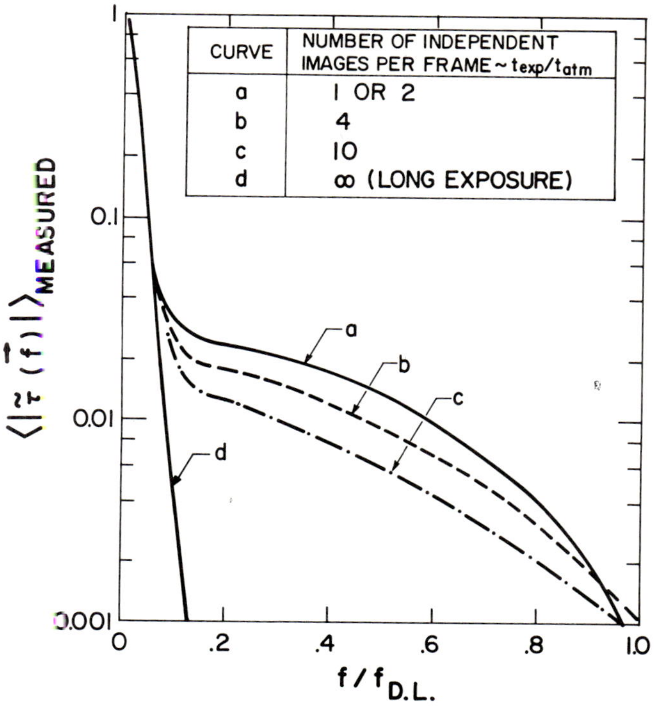

Fig. 2. The effect of exposure time on measurements of $\langle |\tilde{\tau}(\vec{f})| \rangle$ (which is easily related to its square).

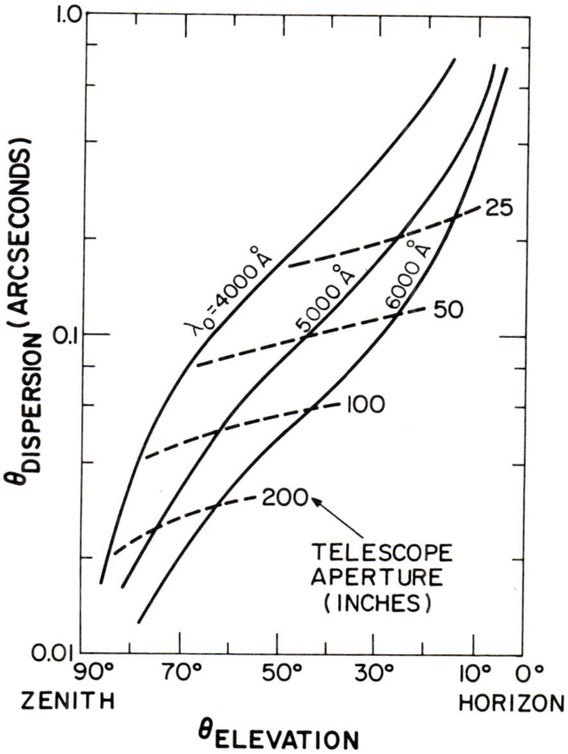

Fig. 3. Atmospheric dispersion as a function of elevation angle for various wavelengths. The dashed curves represent the diffraction limited resolution.

where n is the standard index of refraction and p is the pressure at the observatory altitude H. Although this is only an approximate means of determining $\Theta_{dispersion}$, it is adequate for our present purposes.

Figure 3 shows $\Theta_{dispersion}$ vs $\Theta_{elevation}$ at several values of λ_o for $\Delta\lambda = 250$Å and $H = 10^4$ft. Shown also on this figure is the diffraction angle $\Theta_{diffraction} = 1.22\ \lambda/D$. Considering a 100" telescope, we find that we are dispersion limited, rather than diffraction limited at elevation angles less than 80° at a wavelength of 4000Å. Thus a speckle interferometer operating under these conditions would require dispersion correction or further reduction of the bandwidth below $\Delta\lambda = 250$Å with the subsequent reduction in S/N. The same system operating at 6000Å would be dispersion limited only below 40° elevation angle. Since the seeing (as well as sky background) is usually poor at this low elevation angle, making observation undesirable, a dispersion corrector could be unnecessary. Also, since dispersion affects only one component of $\vec{f}$, there may be cases where useful information can be obtained normal to the dispersion direction, thus making correction unnecessary.

The nominal direction of dispersion at any point in the sky is toward zenith. In the absence of firm evidence to the contrary, some field experience leads us to suspect that both the direction and strength of atmospheric dispersion are random variables of unknown variance, and time constant. Both the direction and level of dispersion, however, can be determined from visual inspection of the magnified white light images, thus providing information on both quantities.

The most obvious means of correcting for a given level of dispersion is through the use of a variable, compensated dispersive prism designed on the basis of Eq. (11). Variability can be obtained with two counter rotating prisms which are in contact with an index matching fluid (i. e., vary $\Theta_{refraction}$) or by the use of a gas filled prism where the particular gas and operating pressure are chosen to compensate for $\Theta_{dispersion}$ over the desired range.

Labeyrie[5] has developed an ingenious means of simultaneously and adjustably compensating for dispersion and selecting λ_o and $\Delta\lambda$ using a special grating. It is this system which we have used extensively in our own work. We shall describe it in detail in the next subsection.

The wavelength and bandwidth selector

Considerations of temporal coherence are usually limited to laser systems, however, the concept of complex degree of coherence, as expressed in the van Cittert-Zernike theorem, is of paramount importance in conventional white light imaging systems. It is this quantity which carries all object information and it is this quantity that the atmosphere unfortunately perturbs. In the absence of the atmosphere, a perfect imaging element (lens or mirror) maintains constant total optical path length. That is, all portions of the wave that arrive at the aperture at a given instant, arrive at the focal plane simultaneously and can therefore properly interfere to form an image. The requirement of maintaining the degree of coherence present in the incident wave does not require a perfect system. This consideration is met over a range of optical path length variations through the imaging system such that the rms path length difference, ΔL, satisfies the relation $\Delta L \ll \lambda_o^2/\Delta\lambda$, the longitudinal coherence length of the light source. In white light, $\lambda_o^2/\Delta\lambda \cong \lambda_o$ which is the reason that good optical systems are figured to better than 1λ and more usually to $\lambda/10$ or $\lambda/20$. However, the lens-atmosphere combination is very poorly figured since the rms phase aberration produced by the atmosphere $\sigma_{\Phi A}$, generally corresponds to a ΔL greater than 1λ. To compensate for this effect, we must require that

$$\Delta L = \frac{\sigma_{\Phi A}}{2\pi} \lambda_o \ll \lambda_o^2/\Delta\lambda \qquad (12)$$

or

$$\frac{\Delta\lambda}{\lambda_o} \ll \frac{2\pi}{\sigma_{\Phi A}} = \text{Order (1)} \qquad (13)$$

By analogy to "good" conventional lenses we therefore require that

$$\frac{\Delta\lambda}{\lambda_o} \sim \frac{1}{10} \text{ to } \frac{1}{20}$$

Thus, with $\lambda_o = 5000Å$, $\Delta\lambda \cong 250\text{-}500Å$. It is apparent that in addition to the loss of coherence, too large a value of $\Delta\lambda$ will produce diffraction blurring even in a conventional imaging system. However, it is this limit which we should push in order to maximize the signal level. In operating a speckle interferometer, it seems advisable to start at a value of $\Delta\lambda$ given above and increase it until diffraction blurring becomes apparent in $\langle|\tilde{\tau}(\vec{f})|^2\rangle$.

Both λ_o and $\Delta\lambda$ can be chosen by use of an interference filter and varied by changing this filter. Some problems have been experienced in practice with this method but the cause is undetermined.

As mentioned in the previous section, Labeyrie has developed a single method for varying λ_o, $\Delta\lambda$ and correction for atmospheric dispersion. This is shown schematically in Figure 4. The magnified image is placed on a figured grating which disperses light in the direction of atmospheric dispersion. This alignment is done with an image rotator (k mirror, Dove prism, etc.) or by rotation of the entire instrument. The figured grating, or a lens used in conjunction with a plane grating, reimages the aperture plane of the telescope, which appears through the magnifying lens at A_1, on a parabolic mirror at A_2 at nearly 1:1 magnification. This parabolic mirror (operating near 1:1) reimages the grating on the sensor. By placing an adjustable aperture at A_2, a variable portion ($\Delta\lambda$) of the dispersed aperture can be selected. By rotating the grating λ_o can be chosen. The dispersion correction ability of this system can best be seen by tracing dispersed red and blue rays through the system. It will be found that these images overlap at I_3 when there is no atmospheric dispersion. The introduction of atmospheric dispersion moves the position of overlap along the optical axis. By proper choice of parameters this movement can be kept within the depth of focus of the system, $\lambda(MF)^2$ where M is the overall magnification. By moving the sensor within the depth of focus, a point can be found where the displaced, colored speckle patterns overlap. At this point, the atmospheric dispersion is compensated.

If a lens is used in conjunction with a flat grating it must be displaced from the grating so as not to interfere with the incident light. This introduces significant aberration into the system. Labeyrie, working with J-Y Optical Systems, Inc. has developed an efficient, figured, aberration corrected grating for this system. It can be obtained commercially from J-Y Optics.

Since the grating is designed to operate in the first order, the zero order (specular) "waste" light is available for continuous visual monitoring. This provides a convenient station for observation of the magnified image and for alignment of the dispersion direction. The grating frequency is chosen so as to provide an adequate dispersion at A_2 and to accommodate the expected atmospheric dispersion within the depth of focus.

Assuming that the dispersion angle is small and the usable F/number of the parabolic mirror (Fig. 4) is large, it can be shown that the maximum correctable dispersion is

$$\Theta_{\text{dispersion}}\Big]_{\max} \cong \frac{f_o \lambda \, \Delta\lambda \, F_{tel}^2}{f_{mag}} \tag{14}$$

where f_o is the grating frequency (e.g. lines/mm) F_{tel} is the F/number of the telescope itself, and f_{mag} is the focal length of the magnification lens.

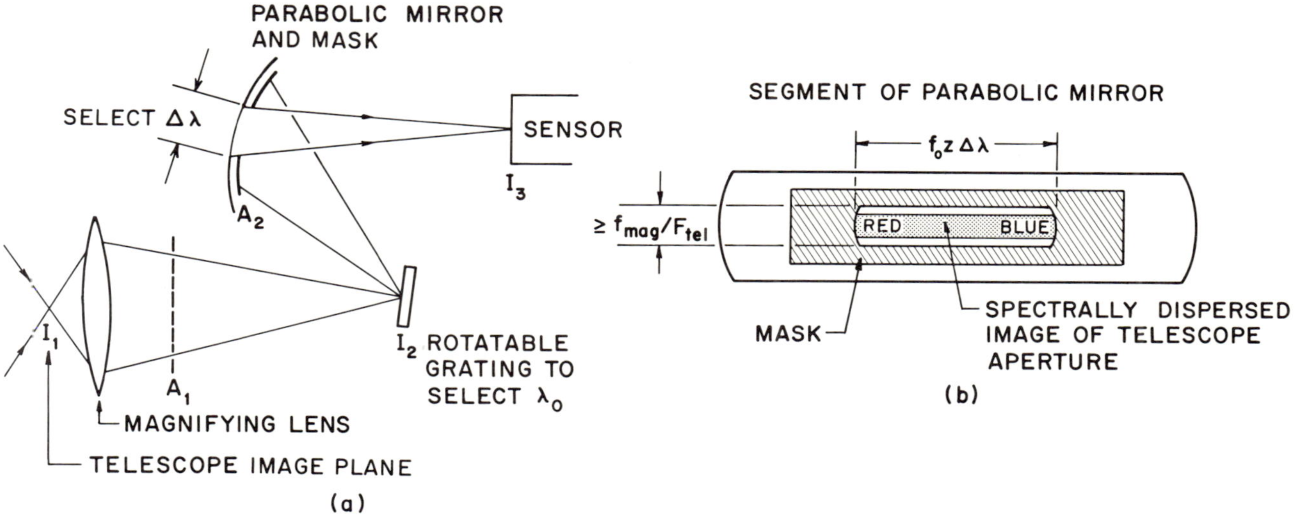

Fig. 4. (a) Schematic drawing of a speckle interferometer optical train. (b) A more detailed view of the parabolic mirror and mask used for varying $\Delta\lambda$.

The mask, placed on the parabolic mirror for wavelength filtering, must have a height (perpendicular to the dispersion direction) of f_{mag}/F_{tel} and a width, w of

$$w \cong f_o Z \Delta\lambda \geq \frac{f_{mag}}{F_{tel}} \tag{15}$$

where Z is the distance from the grating to the mirror.

Labeyrie's grating, with a frequency of 1800 ℓ/mm and a design Z = 272 mm, will correct over 25 $\widehat{\text{II}}$ of dispersion for our nominal system. The required mask is 1.6 mm x 14 mm for $\Delta\lambda$ = 250Å. Although a lower value of $\Theta_{\text{dispersion}}\Big]_{\max}$ would have been more than adequate (see Fig. 3) it would have compromised the wavelength filtering requirement. This system clearly has a precision tradeoff between dispersion correction and $\Delta\lambda$ selection.

The Sensor

Most of the requirements for a sensor were described in previous sections where we determined the expected number of photons/speckle and stated the required number of pixels per $\lambda_o F$. If we consider that the speckle interferometer has an overall transmission, T, and a quantum efficiency, η, we find that the number of photo events per pixel for a point object is given by

$$\mathcal{E} = \frac{n_p T\eta}{\alpha^2} \tag{16}$$

or, for our previous example, with T = 0.1, η = 0.2 and α=2, $\mathcal{E} \cong$ 10 photoevents/pixel.

It is clear that we must use a "photon limited" sensor for most realistic SI applications. A decade ago, this meant a multistaged image intensifier plus fast film. Today there are a number of pure electronic devices including SIT, SEC, EBS, SEM, ISEC, and ISIT vidicons which have the required sensitivity and resolution. The recent development of CCD's makes them promising candidates for use in speckle interferometers.

It seems safe to say that if the noise equivalent counts per pixel are less than $\sqrt{\mathcal{E}}$, a given sensor should be useful for SI. It is extremely difficult to determine this noise equivalent counts per pixel from vendor spec sheets so we will have to leave the answer to that question to the reader. Experience seems to indicate that the sensor which has the least gain necessary to provide photon limited operation at the expected photons per pixel is the best choice. Higher sensitivity often leads to unnecessary distortion in $|O(\vec{f})|$.

Current efforts in Europe using multistage image intensification are directed at determining the spacing of 14th magnitude binaries. It is doubtful that current detectors will permit more than the crudest information on more complex objects of comparable brightness.

Data Recording and Analysis

Without dwelling extensively on this subject we can note a few relevant points.

The inherent data rate for an unresolved object is approximately

$$\frac{(D/r_o)^2 \; \alpha^2}{t_{atm}} \quad \log_2 \mathcal{E} \; \frac{\text{bits}}{\text{sec}} \tag{17}$$

assuming that the x-y coordinates are mapped into time through the scan rate and scan pattern. In our previous example this amounts to approximately 10^6 bits/sec, a rate consistent with many high quality magnetic tape recorders. Imaging extended objects may increase this rate by a factor of 10.

Alternate coding techniques can reduce this load substantially while too large a FOV increases it unnecessarily.

This rate will normally be moderated by the sensor's available duty cycle, since framing at t_{atm}^{-1} (as large as 10^3) without image persistance (memory of previous frames) is beyond the present state of the art. Most current sensors have duty cycles of from .5 to .01, depending on exposure time.

By appropriate buffering, FOV control and data encoding, recording in either analog or digital form should not be difficult.

Both large size general purpose computers and small specialized hardware FFT devices can Fourier transform data at rates in excess of 10^5 points/sec. Thus images of 128 x 128 pixels can be processed at the rate of ~ 10/sec. This rate is close to the effective framing rate of current commercial sensors so that a near real time digital data processing (and therefore data compression) capability currently exists.

Work on analog optical(9) and electrical processing may lead to real time analog capabilities with resulting effective substantial reduction of recording requirements. However, real time processing eliminates the option of reprocessing the data itself. This can be a substantial disadvantage for all but the most routine applications of SI.

Before we turn to the question of S/N we note the following for completeness (see Fig. 5).

1. The speckle interferometer structure should be both rigid and lightweight. Time spent in the careful design and construction of the structure will be rewarded in the precious minutes of mount time that are spent collecting data rather than realigning a "rubber" optical system. The inclusion in the instrument of a small alignment laser will greatly aid initial adjustment of the system and provides an easy means of checking performance when the instrument is mounted on the telescope. Also, a back illuminated target (Ronchi ruling or fine scale speckle negative) placed at the telescope focal plane can be used to calibrate the overall performance of the entire speckle interferometer.

2. Hands on operation of a speckle interferometer is essential. By placing viewing stations (i.e., folding mirrors and eyepieces at judicious places in the optical system the various knobs on the instrument (λ_o, $\Delta\lambda$, dispersion correction, t_{exp}, etc.) can be properly adjusted. Remote monitoring and mechanical actuators are complex and expensive alternatives in most astronomical applications.

3. L^3 sensors and associated optical elements require suitable light baffling.

4. A speckle interferometer, since it operates at the highest magnifications, often shows up previously unknown telescope characteristics such as mount jitter and mistrack along with aberrations previously masked by seeing. Although SI is insensitive to most aberrations(10), they can affect S/N by reducing the number of photoevents per speckle (see below).

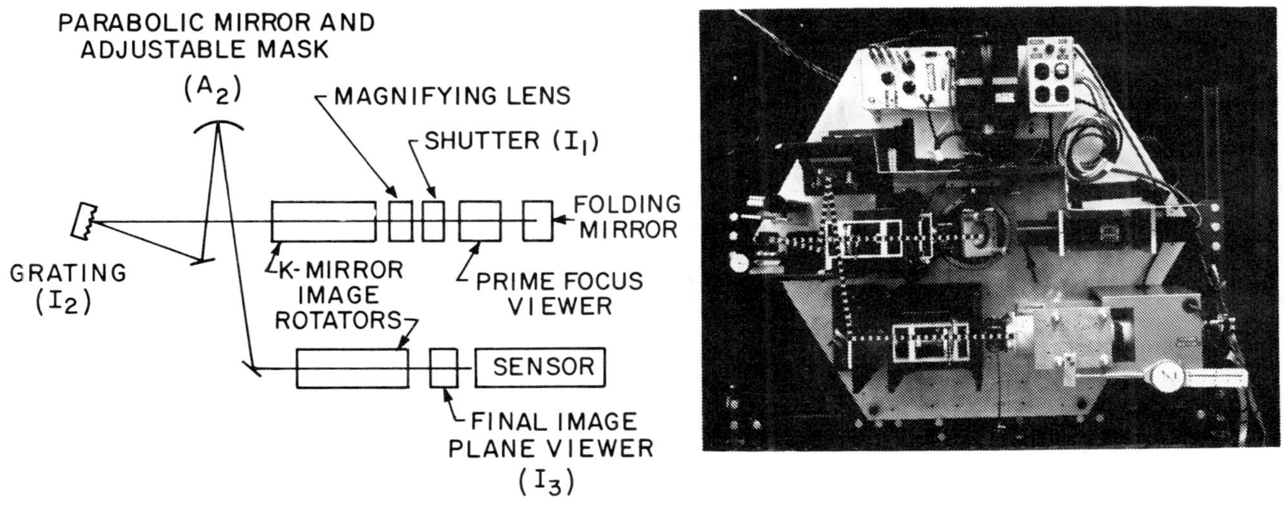

Fig. 5. A speckle interferometer used for measurement of atmospheric parameters that affect SI (cover removed).

SI Signal to Noise Ratio

After having established the design criteria for a speckle interferometer, we can now determine the applicability of SI to a given imaging problem. We assume that knowledge of $|O(\vec{f})|$ has been determined to be useful and ask how accurately we can measure this quantity using SI. For this purpose we must first determine the accuracy to which we can measure $\langle|\tilde{\tau}(\vec{f})|^2\rangle$.

Figure 6 represents a qualitatively accurate plot of this quantity. Figure 6a shows the mathematical limiting shape. It is clear that, in principle and for "well behaved noise", i. e., white noise $\langle|\tilde{\tau}(\vec{f})|^2\rangle$ can be extracted from the noise with arbitrary accuracy. In practice, however, the limited number of ensemble members (frames), n, used in obtaining the estimate of $\langle|\tilde{\tau}(\vec{f})|^2\rangle$ will produce a total variance, σ_T^2, in the random components caused by speckle and noise, both photon and detector, (Fig. 6b). Because our primary interest is $\langle|\tilde{\tau}(\vec{f})|^2\rangle$ for frequencies $f/f_{D.L.}$ between r_o/D and unity in Fig. 6a, we can conveniently define a signal to noise ratio

$$\frac{S}{N} \equiv \frac{①}{\sigma_T} \tag{18}$$

Since the individual frames of data are statistically independent, $\sigma_T = (1/\sqrt{n})\,\sigma_{Ti}$ where σ_{Ti}^2 is the variance of a single ensemble member, $|\tilde{\tau}(\vec{f})|^2$. By the central limit theorem, it can be shown that even in the absence of other noise sources, statistical considerations alone produce a variance in the power spectrum equal to its expected value, i. e. $\sigma_{Ti} = \langle|\tilde{\tau}(\vec{f})|^2\rangle + \sigma_{Ti}(\text{noise})$.

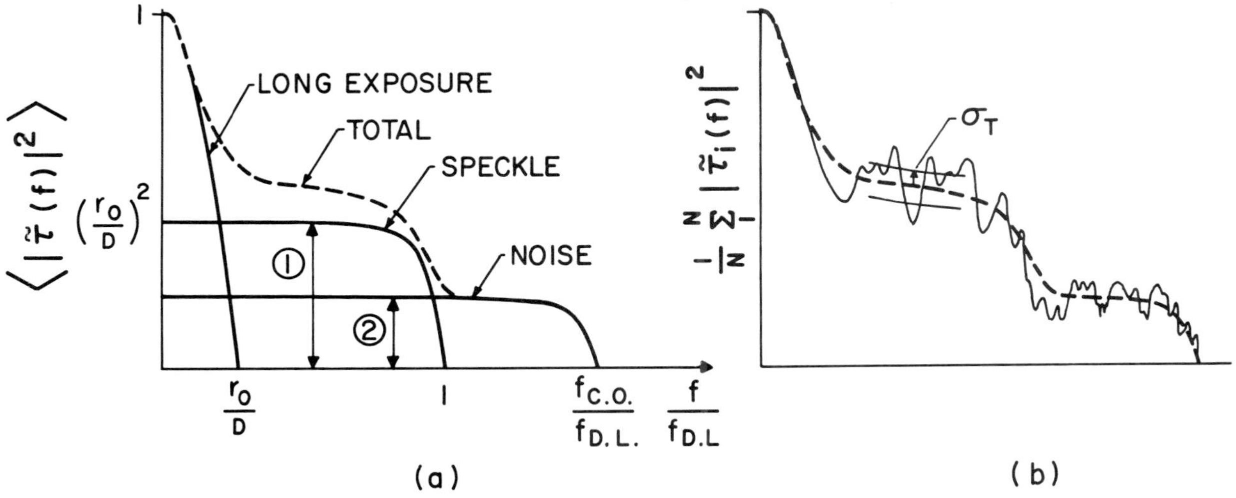

Fig. 6. A model of $\langle|\tilde{\tau}(\vec{f})|^2\rangle$ for calculation of S/N, (a) theory, (b) typical measurement.

Therefore, we can write

$$\frac{S}{N} = \frac{\sqrt{n}}{1 + \sigma_{Ti}(\text{noise})/\langle|\tilde{\tau}(\vec{f})|^2\rangle} \tag{19}$$

From this result, it is immediately apparent that $\frac{S}{N} \leq \sqrt{n}$, i.e., a large number of data frames are required to produce high S/N.

Noise arises from both photon statistics and the detector itself. Since the latter source is so dependent on the nature of the sensor it is difficult to treat it in a general way. Little sensor data exists which can be used to determine the appropriate contribution of detector noise variance. We shall therefore consider only the contribution of photo statistics to the noise variance. There are several high quality L^3 sensors which make this approximation appropriate for moderately bright objects.

By noting that the photon noise in a given speckle is proportional to $\sqrt{n_p}$, it can be shown that photon noise contribution to Eq. (19) is $1/\mathcal{E}\mathfrak{a}^2$ or

$$\frac{S}{N} = \frac{\sqrt{n}}{1 + 1/\mathcal{E}\mathfrak{a}^2} \tag{20}$$

for a photon limited speckle interferometer. Thus, in our previous example where $\mathcal{E}\mathfrak{a}^2 \cong 40$, $S/N \cong \sqrt{n}$; i.e., the signal to noise ratio is frame limited. For a m = 14 star, $\mathcal{E}\mathfrak{a}^2 \cong 10^{-2}$ detected photoevents per speckle so that $S/N \cong 10^{-2}\sqrt{n}$. Thus a S/N = 1 requires at least 10^4 frames of data. To resolve a m = 14 binary using SI would probably require 10^6 frames since some censoring of the data would be required to eliminate frames having an insufficient number of detected events. At a framing rate of 10/sec, this would require over 24 hrs. of data collection or about a week's observation time.

To determine the effect of this S/N in the case of an extended object, we must return to Eq. (1). For example, consider the donut shaped object of angular size $\underline{S}$ which has a hole of angular size $a\underline{S}$, then

$$|O(f')|^2 \sim \left(\frac{J_1(f')}{f'} - a^2\frac{J_1(af')}{af'}\right)^2 \tag{21}$$

that is, information about the hole is contained in $|O(f')|^2$ at a level of approximately a^2 or lower. (It is for this reason that moderate central obscurations in telescopes do not appreciably affect the point spread function.)

When we rewrite equation (1) in the form

$$|\tilde{I}(\vec{f})|^2 = |O'(\vec{f})|^2 \; |\tilde{\tau}(\vec{f})|^2 \times (1 + a^2 O_a(\vec{f}) + \ldots)\;(1 + N/S) \tag{22}$$

we immediately see that information about the small detail (of order a^2) can be masked in the noise of the large detail (of order N/S) or, in other words, we must require that for an extended object

$$S/N \gg 1/a^2 \tag{23}$$

where $a \cong$ finest resolvable object detail/object size.

To see diffraction limited information on a 1 $''$ object we require, under our previous conditions, $S/N > 10^2$. From Eq. (20) we can easily determine the required number of frames for a given object brightness/diffraction limited element.

In general, the information of this section can be used to estimate the number of data frames required for a given measurement goal. Note that regardless of how small $\mathcal{E}\mathfrak{a}^2$ is, a sufficiently large number of frames will produce a usable S/N. The requirement of temporal stationarity, however, limits the number of realizations that can be obtained and analyzed at constant $\langle|\tilde{\tau}(\vec{f})|^2\rangle$. By co-adding the reduced results obtained during many shorter periods, the higher S/N can be synthesized.

Summary and Conclusion

We have attempted in this paper to review the principles of SI, characterize the speckle images themselves, propose design considerations for the construction of a speckle interferometer and determine its usefulness (i.e. S/N) in a given application.

Lest we frighten any potential user by our desire to be complete in addressing these subjects, we wish to point out that the initial speckle interferometer built by Labeyrie and his co-workers, which satisfied all of the above criteria (for bright, simple objects) probably cost under $5000, the chief cost being a motorized 35 mm camera which was used to record the 200 frames of image intensitifed photographic data.

The resulting data was processed on an optical Fourier transform bench which, aside from a He Ne laser and a Polaroid film holder must have cost all of $25.

Ingenuity combined with an understanding of SI can inexpensively open up the resolution range to anyone using a telescope of 10 cm or greater aperture and allow him to "beat the atmosphere".

Recurrent Nomenclature

D	telescope aperture diameter
F, F_{tel}	F number (of telescope) = $\mathscr{f}/D$
FOV	field of view
$\tilde{I}(\vec{f})$	Fourier transform of $\tilde{I}(\vec{x})$
$\tilde{I}(\vec{x})$	speckle image irradiance (p for point object)
K-T	Knox-Thompson (phase retrieval algorithm)
L^3	Low Light Level
M	magnification factor in speckle interferometer (Eq. (9))
MTF	modulation transfer function
$O(\vec{x})$	perfect image (diffraction limited)
$O(\vec{f})$	transform of $O(\vec{x})$
S/N	signal to noise ratio
$\underline{S}$	angular size of resolvable object
SI	Speckle Interferometry
$\vec{f}$	spatial frequency in Fourier transform plane
$f_{D.L.}$	diffraction limited spatial frequency $D/\lambda\mathscr{f}$
f_s	seeing limited spatial frequency = $r_o/\lambda\mathscr{f}$
m, m_i	apparent photometric magnitude
n	number of independent speckle images processed
r_o	correlation scale of the atmospherically induced wavefront aberrations at the aperture (see reference 13)
t_{atm}	atmospheric time constant
t_{exp}	exposure time
α	detector sampling ratio (pixels per λ F)
$\mathscr{n}_p$	number of photons per speckle for a point object (Eq. (8))
$\mathscr{f}$	basic telescope focal length
$\mathcal{E}$	number of photo events per pixel for a point object (Eq. (16))
$\tilde{\Phi}(\vec{f})$	phase of $\tilde{I}(\vec{f})$
$\lambda, \lambda_o, \Delta\lambda$	wavelength (mean and band width)
$\tilde{\tau}(\vec{f}), \tilde{\tau}_i(\vec{f})$	instantaneous optical transfer function for the lens-atmosphere combination (from source i)
$\|(---)\|$	modulus
$\langle(----)\rangle$	ensemble average

$\langle(----)\rangle_m$ measured average

$(\tilde{-})$ random variable

References

1. A. Labeyrie, Astron. Astrophys. 6, 85 (1970).
2. D. Y. Gezari, A. Labeyrie, and R. V. Stachnik, Astrophys. J. (Letters) 173, Ll (1972).
3. D. Bonneau and A. Labeyrie, Astrophys. J. (Letters) 181, Li (1973).
4. A. Labeyrie, D. Bonneau, R. V. Stachnik, and D. Y. Gezari, Astrophys. J. (Letters) 194, L147 (1974).
5. A. Labeyrie, Nouvelle Revue d'Optique 5, 141 (1974).
6. D. Korff, G. Dryden, and R. P. Leavitt, JOSA 65, 1321 (1975).
7. A. M. Schneiderman, P. F. Kellen and M. G. Miller, JOSA 65, 1287 (1975).
8. D. P. Karo and A. M. Schneiderman, Topical Meeting on Speckle Phenomena, Th C6, Pacific Grove, Calif. (1976).
9. P. Nisenson and R. V. Stachnik, Imaging in Astronomy technical digest, Th C2, Cambridge, Mass. (1975).
10. D. Korff, JOSA 63, 971 (1973).
11. K. T. Knox and B. J. Thompson, Astrophys. J. (Letters) 193, L45 (1974).
12. R. H. T. Bates, P. T. Gough, and P. J. Napier, Astro. and Ap. 22, 319 (1973).
13. D. L. Fried, JOSA 56, 1372 (1966).
14. C. Roddier and F. Roddier, JOSA 65, 664 (1975).

ASTRONOMICAL SPECKLE IMAGING

P. Nisenson, D. C. Ehn and R. V. Stachnik
Itek Corporation, Optical Systems Division
10 Maguire Road, Lexington, Massachusetts 02173

Abstract

Speckle imaging is a technique for recovering diffraction limited images from sequences of atmosphere-degraded, short exposure photographs obtained at a large telescope. The technique is derived from speckle interferometry and shares many of the characteristics of that process, including dependence of the output signal-to-noise on number of frames processed and relative insensitivity to fixed telescope aberrations and noise in the image record. Speckle interferometry has been demonstrated to yield telescope-diffraction-limited information, but only in the form of spatial power spectra. Speckle imaging averages a different quantity, the statistical autocorrelation of the image Fourier transform, which contains all the information in the averaged power spectra plus the transform phase information required to recover an image. Two-dimensional digital simulations of the process for extended continuous-tone objects are presented, and include the case where severe static telescope aberrations are present.

Introduction

Speckle imaging is a technique for recovering diffraction-limited resolution from atmospherically degraded images recorded at the focus of large up-looking telescopes. This technique has evolved from speckle interferometry, a process that has been extensively analyzed and proven in the field yielding many astronomically significant results. However, in speckle interferometry, only image autocorrelations are obtained from the process, while in speckle imaging, true diffraction-limited images can be reconstructed. In both processes, a series of short-exposure, highly magnified images are recorded, allowing sufficient time between exposures so that the atmosphere is decorrelated from frame to frame. In speckle interferometry, the recorded images are generally optically Fourier transformed and then individual power spectra from each image are integrated together. This procedure is equivalent to summing the autocorrelations of each speckle image. The process has been demonstrated to yield diffraction-limited results with the signal-to-noise ratio proportional to the square root of the number of frames integrated together. In speckle imaging, the averaging process is nearly identical with the interferometry process except that the phase in each Fourier transform is encoded in such a way that, after averaging, both the object amplitude and phase can be recovered and true images reconstructed. This process is based on a technique suggested by Knox and Thompson (1974) and developed and demonstrated at Itek (Nisenson and Ehn, 1974) using computer simulations for extended, continuous tone objects. More recently, an optical processing system for analog implementation of the averaging step in the speckle imaging process has been designed and tested in the laboratory. Preliminary analysis indicates that the process works well even at very low light levels and can recover high resolution images in the presence of severe static telescope aberrations and with limited signal-to-noise ratio in the recording process.

Speckle Interferometry

Speckle interferometry is a technique suggested by Labeyrie (1970) for the recovery of diffraction-limited information from short exposure images of astronomical objects. Laberyrie's approach to recovering this information was to add together the spatial power spectra of the images rather than the images themselves. This procedure preserves information that is lost in the more conventional long-exposure averaging process. Fig. 1 illustrates the Labeyrie approach.

The power spectra at the bottom of the figure show clear evidence of image information content well beyond the long-exposure seeing limit (here, about 2 arc-seconds). The star on the left, Betelgeuse, has an angular diameter of 0.05 arc-second and shows enlarged speckle and a small power spectrum disk. Capella, in the center, is a binary star having a separation between components of 0.05 arc-seconds. It shows double speckle and a fringe pattern in the power spectrum. Finally, Vega, which is unresolved at the diffraction limit, shows sharp speckles and a large power spectrum disk. Adding together power spectra improves the signal-to-noise ratio of the result. Retransformation of the power spectra would yield near diffraction limited autocorrelation images.

Speckle interferometry has been used by Gezari, Labeyrie, and Stachnik (1972) to measure stellar diameters; by Labeyrie, Bonneau, Stachnik, and Gezari (1974) to observe close binaries; by Bonneau and Labeyrie (1973) to observe possible stellar surface features; and by Harvey and Breckinridge (1973) and Schwartzchild and Harvey (1975) to observe solar surface features. These observations demonstrated that diffraction-limited information can be recovered for realistic astronomical sources.

The observations also demonstrate the fact that the telescope need not be diffraction limited to produce diffraction-limited information. Fig. 2 shows a Gaussian fit to the measured point spread function of the 200-inch telescope and, superimposed on it, the calculated diffraction-limited spread function.

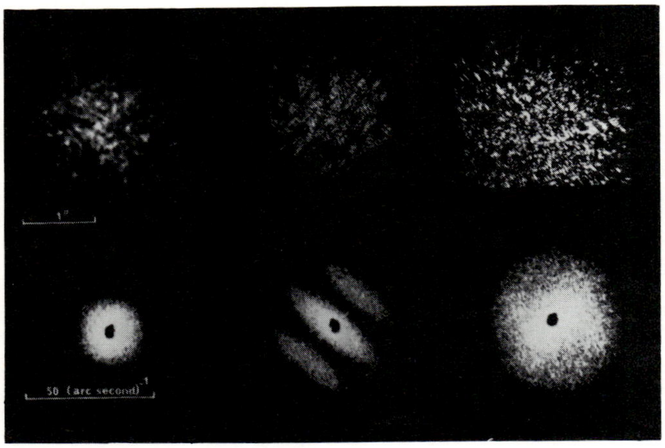

Fig. 1 — The top row shows direct, highly magnified short-exposure images of the stars Betelgeuse, Capella, and Vega. The bottom row shows spatial power spectra produced from 50 images like those in the top row. The power spectra demonstrate that Betelgeuse is a resolved star with an angular diameter of 0.05 arc-second, Capella is a binary star with a separation of 0.05 arc-second, and Vega is unresolved at the diffraction limit (0.025 arc-second) of the 200-inch telescope (Labeyrie, Bonneau, Stachnik, and Gezari, 1974).

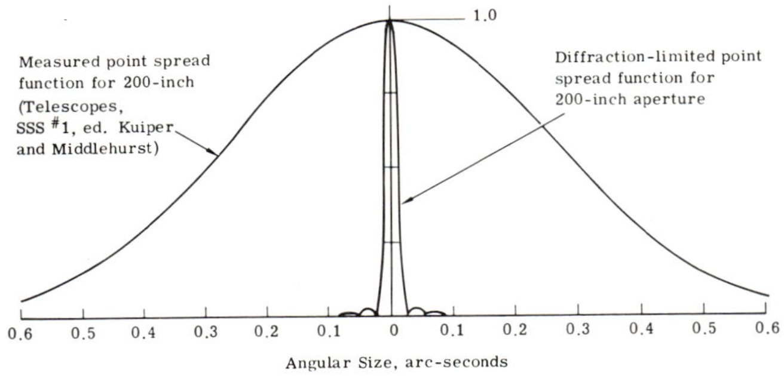

Fig. 2 — Point spread function of a diffraction-limited 200-inch aperture and that of the 200-inch Palomar telescope as determined from the measured aberrations.

The corresponding modulation transfer functions (MTF) implied by these curves are plotted at the top of Fig. 3. Plotted below the transfer functions, on the same horizontal (spatial frequency) scale, is the intensity of the fringe pattern measured for the binary star Capella. Evidently, no fringes at all would be observed if the transfer function of the speckle process were not substantially better than the long exposure transfer function. The fringe pattern, in fact, indicates that the speckle MTF is essentially diffraction limited. This is borne out by calculation of the effect of fixed telescope aberrations on the speckle transfer function $\langle |T|^2 \rangle^{1/2}$ where T is the instantaneous transfer function, and $\langle\,\rangle$ represents a time average.

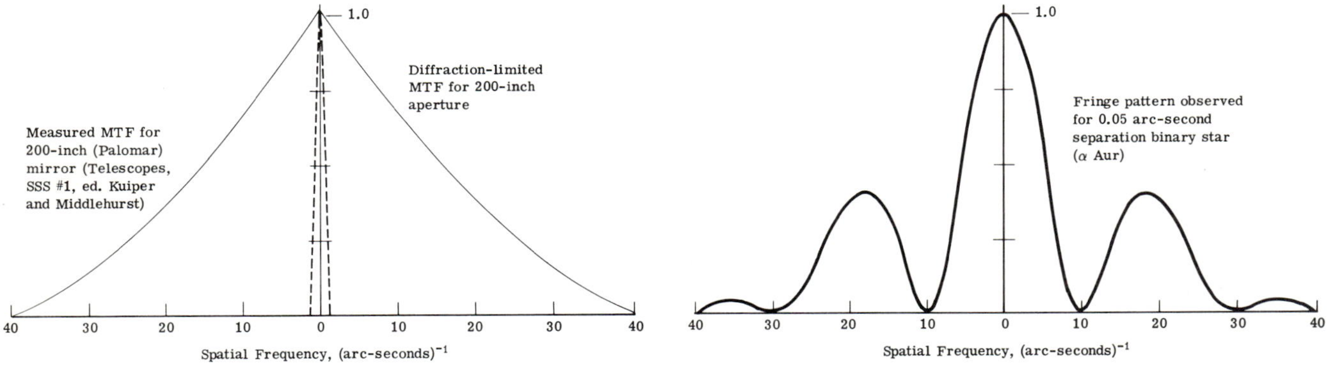

Fig. 3 — The transfer functions at the top of the page are those expected for a diffraction-limited 200-inch aperture and for the Palomar 200-inch telescope on the basis of its measured aberrations. The fringe pattern observed for the binary star Capella (α Aur) is plotted at the bottom on the same spatial frequency scale. It is evident that the transfer function of the speckle process is essentially diffraction limited even in the presence of substantial telescope aberration.

Curves showing the reduction in the transfer function due to aberrations appear in Fig. 4. They show the ratio between the aberrated and unaberrated cases evaluated for a one-dimensional, 1-meter telescope and for an atmosphere correlation length, r_0, of 0.1 meter. For such a telescope and atmosphere, the wavefront error arising from the Kolmolgorov turbulence spectrum of the atmosphere alone will have an rms value of 1.1 waves. Note that the reduction in the speckle transfer function is still tolerable for fixed aberrations significantly larger than the aberrations associated with the random wavefront. Although a rigorous analysis of the speckle imaging transfer function has not yet been carried out, the interferometry and imaging transfer functions should be nearly identical, including insensitivity to aberrations.

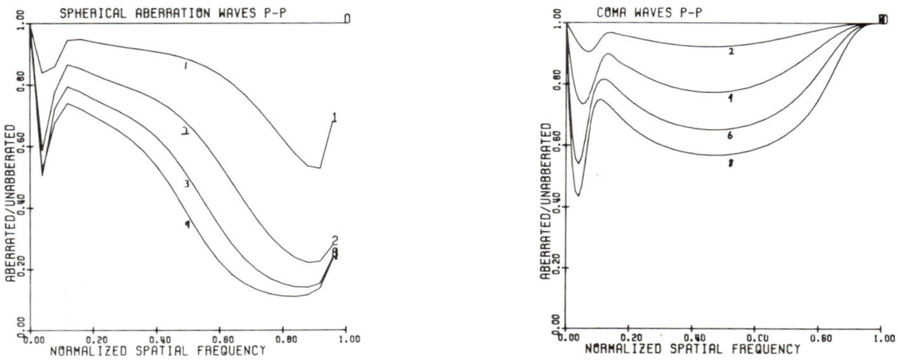

Fig. 4 — Curves showing the degradation of the speckle transfer function due to coma and spherical aberration. In each case, the vertical axis corresponds to the ratio of the aberrated speckle transfer function to the unaberrated transfer function of a 1-meter telescope in 1-arc-second seeing. The horizontal axis is normalized spatial frequency. The individual curves are labeled with the number of waves of aberration present.

Speckle Imaging

A serious shortcoming of speckle interferometry is that spatial power spectra or autocorrelation images are obtained from the process instead of true images. This result is only easily interpretable for very simple objects and actual images can be reconstructed only for centrosymmetric objects.

In speckle interferometry, the short exposure speckle images are Fourier transformed and their corresponding power spectra integrated, usually in a coherent optical processor. The averaged intensity in the transform, $\langle|I(u,v)|^2\rangle$ takes the form

$$\langle|\tilde{I}(u,v)|^2\rangle = |\tilde{O}(u,v)|^2 \langle|\tilde{T}(u,v)|^2\rangle \qquad (1)$$

where $\tilde{O}(u,v)$ is the object spectrum and $\tilde{T}(u,v)$ is the atmospheric and optics transfer function, and u and v are the coordinates in the frequency plane. While it has been shown that the transfer function of the process yields information out to the diffraction limit, the phase in the object transform is lost and only the autocorrelation of the object's image can be reconstructed.

In a recent paper (K. T. Knox and B. J. Thompson, Astrophys. J. Letters 193, L45 (1974)), Knox and Thompson (K&T) discuss an algorithm for digital processing of astronomical speckle data to obtain diffraction-limited images. They demonstrate that in their processing of speckle images, the phase in the averaged object transform can be preserved, allowing reconstruction of true images. At the same time, K&T have shown that the transfer function of their process is almost exactly equivalent to the transfer function of the standard speckle process. K&T require digitization of each speckle image, followed by two-dimensional Fourier transformation of each image.

In their paper, K&T suggest averaging of the statistical autocorrelation, $\tilde{A}(u,v)$, of the image transform instead of the power spectrum. The averaged quantity $(\tilde{A}(u,v))$ takes the form:

$$\begin{aligned} \langle\tilde{A}_u(u,v)\rangle &= \langle\tilde{I}(u,v)\,\tilde{I}(u+\Delta u, v)\rangle \\ \langle\tilde{A}_v(u,v)\rangle &= \langle\tilde{I}(u,v)\,\tilde{I}(u, v+\Delta v)\rangle \end{aligned} \qquad (2)$$

Δu and Δv are required to yield small shifts compared to the correlation distance of the speckle transform. The two separate averages encode the phase of the transform in the u and v direction respectively.

This can be seen when equation (2) is expanded (only the "u" term is considered though the result is equivalent for the "v" term).

$$\left\langle \tilde{A}_u(u, v)\right\rangle = |\tilde{O}(u, v)| e^{i\phi(u, v)} \cdot |\tilde{O}(u + \Delta u, v)|\; e^{-i\phi(u + \Delta u, v)} \cdot \Big\langle |\tilde{T}(u, v)|\, e^{i\psi(u, v)} \cdot |\tilde{T}(u + \Delta u, v)| e^{-i\psi(u + \Delta u, v)} \Big\rangle \tag{3}$$

where $\phi(u, v)$ and $\psi(u, v)$ are the object and transfer function phase distributions respectively. This can be rewritten as

$$\left\langle \tilde{A}_u(u, v)\right\rangle \simeq |\tilde{O}(u, v)|^2\, e^{i\,\Delta\phi(u, v)} \cdot \Big\langle |\tilde{T}(u, v)|^2\, e^{i\,\Delta\psi(u, v)} \Big\rangle \tag{4}$$

where

$$\Delta\phi\;(u, v) = \phi(u, v) - \phi(u + \Delta u, v)$$

and

$$\Delta\psi\;(u, v) = \psi(u, v) - \psi(u + \Delta u, v)$$

This process has encoded the phase in the transform as phase differences between adjacent correlated points in the transform. The transfer function term in Eq. (4) has an expected value whose phase approaches zero and whose amplitude approaches the diffraction limited transfer function usually obtained with conventional speckle interferometry. After averaging, the point to point phase differences are known for a grid of points in the object transform. The relative phase at any point in the transform is then calculated by summing the phase differences between the point and an arbitrary reference point (usually the origin). Averaging the calculated phase for the multiplicity of phase paths between points greatly reduces the effects of noise and the sensitivity to bad data points. Itek is currently developing a number of different algorithms for optimally calculating the phase while minimizing computer time.

A simulation of the speckle imaging process performed digitally appears in Fig. 5. The figure shows, respectively, a diffraction-limited image of the Comsat satellite as seen by a 40-inch telescope, an atmospherically degraded (1-arc-second seeing) short-exposure image of the satellite, the result of adding 40 degraded images directly, and the result of adding 40 degraded images using the speckle process.

Diffraction-limited image of Comsat satellite

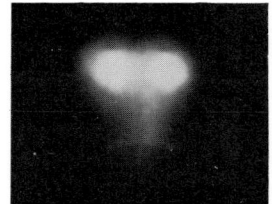

Short exposure image of Comsat with atmospheric degradation comparable to that expected for a 1.5 meter telescope

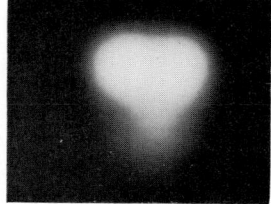

Long exposure image (40 short exposures added directly) with atmospheric degradation

Image recovered from the 40 short exposures using the speckle averaging process

Fig. 5 — Speckle imaging simulation

Fig. 6 shows the results of a simulation investigating the effects of fixed telescope aberrations on the reconstruction. In the upper left is the diffraction-limited image of a "triple star." The two closest components are separated by twice the Rayleigh resolution distance and are half the brightness of the bright component. At the lower left is the reconstruction obtained from 35 frames taken with an unaberrated telescope. At the upper right is the image that would be obtained in the presence of 4 waves (peak-to-peak) of spherical aberration and no atmosphere. Finally, the image at lower right shows the reconstruction obtained from 35 atmosphere-affected frames taken with the aberrated telescope. The image that emerges from speckle processing of the aberrated telescope data is substantially better than that which would result if the telescope could be used in the absence of atmospheric seeing effects.

The insensitivity of both speckle imaging and speckle interferometry to aberrations arises because of the way the high resolution information is coded into the short-exposure image. The criterion for diffraction-limited information recovery in the presence of aberration is that the horizontal scale of the telescope-induced wavefront distortions be larger than the horizontal scale of the atmosphere-induced distortions. This is equivalent to requiring that the telescope be of sufficient quality to resolve the seeing disk. If this circumstance holds, the wavefront may be regarded as being formed by many subapertures (correlation areas) which themselves produce images that overlap at the focal plane of the telescope. Only the aberration over each independent correlation area, rather than the aberration over the full aperture, has any effect on the high resolution image information coded into the speckle carrier pattern. The aberration sensitivity of the speckle process, then, is comparable to that expected from a telescope with its aperture stopped down to the atmosphere correlation distance, typically less than 10 centimeters.

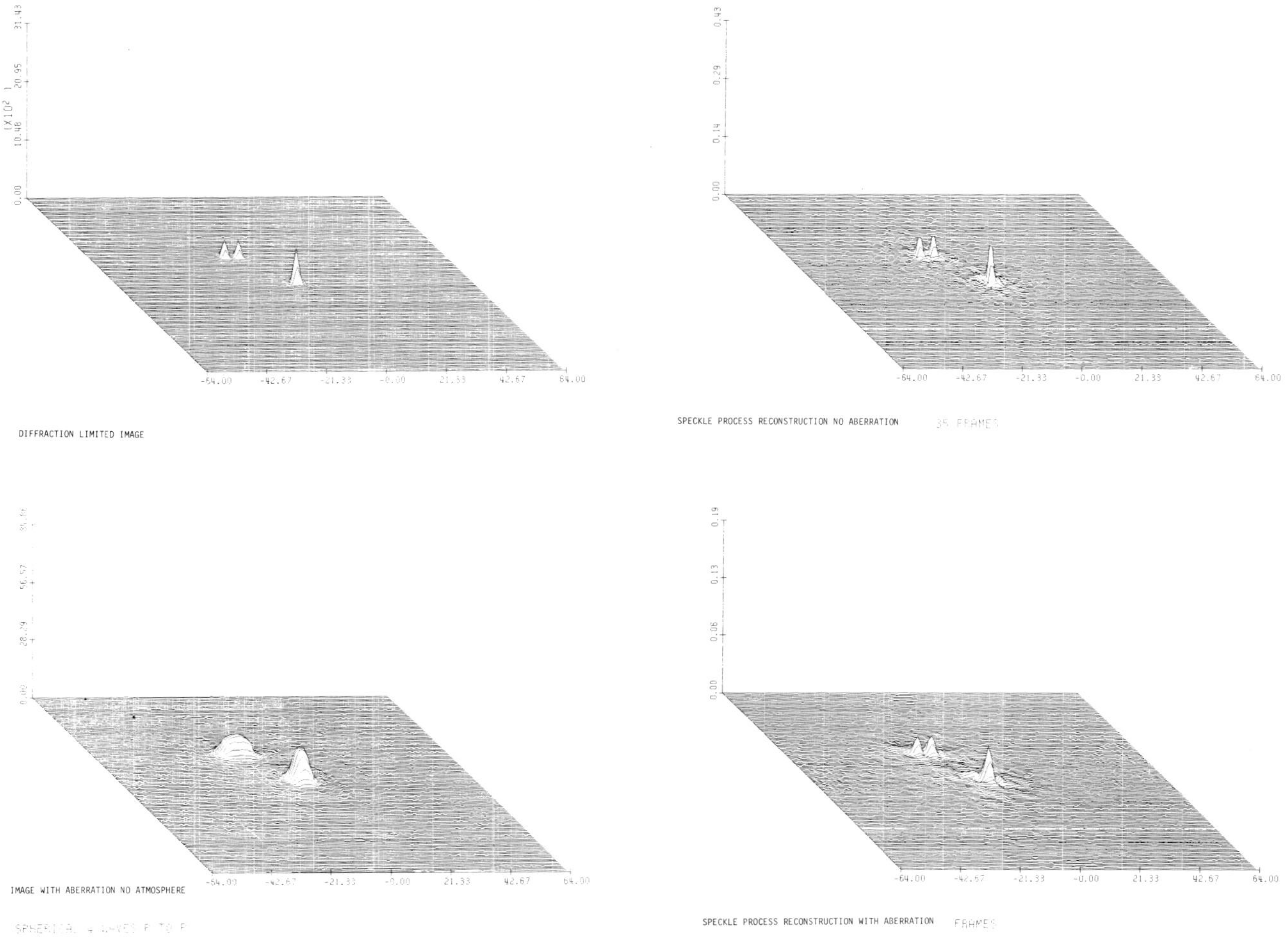

Fig. 6 — Demonstration of the imaging process in the presence of aberrations. At the top left is the image of a simulated triple star as recorded by an unaberrated diffraction-limited telescope. Below it is the image recovered from 35 atmosphere-affected, short-exposure frames of the object. This image is analogous to the final picture in Fig. 5. At the top right is the image that would be recorded by an aberrated telescope in the absence of atmospheric degradation. The aberration used is 4 waves (peak-to-peak) of the spherical. The image at the bottom right is the result of processing 35 atmospherically-degraded frames using speckle imaging. Note that the reconstruction is only weakly affected by the aberrations and is nearly diffraction limited.

Conclusion

The speckle imaging process appears to have an enormous potential in that it could lead to the development of extremely large aperture telescopes or telescope arrays that would yield images with resolution down to their diffraction limit. The results obtained with speckle interferometry at the 200-inch telescope (Gezari et al., 1972, Labeyrie, et al., 1974) and by Labeyrie with a two-telescope system (Labeyrie, 1975) indicate that extending the process to larger apertures or optical arrays is almost certainly practical. The insensitivity of the process to severe static aberrations also indicates that constraints on the quality of the optics could be greatly relaxed and new system designs and support structures could be considered.

Because of its relative simplicity and low cost and because the magnitude limit depends only on the available integration time, speckle imaging is well suited to astronomical applications. Improvements in imaging resolution obtained with balloon-borne experiments and several ground-based imaging techniques, as well as increased resolution from non-visible wavelength imaging systems and photographs returned by planetary probes indicate that even small improvements in resolution can lead to the discovery of new features and phenomena. It appears likely that the very high resolution obtainable with speckle imaging could well play an important role in astronomy in the coming years.

References

1. Boidin, D., and Labeyrie, A., Speckle Interferometry in the Photon-Counting Mode, unpublished report (1973).
2. Bonneau, D., and Labeyrie, A., Astrophysical Journal, 181, p. L1 (1973).
3. Gezari, D. Y., Labeyrie, A., and Stachnik, R. V., Astrophysical Journal, 178, p. L1 (1972).
4. Harvey, J. W., and Breckinridge, J., Astrophysical Journal, 182, p. L137 (1973).
5. Knox, K. T., and Thompson, B. J., Astrophysical Journal, 193, p. L45 (1974).
6. Labeyrie, A., Bonneau, D., Stachnik, R. V., and Gezari, D. Y., Astrophysical Journal, 194, p. 147 (1974).
7. Nisenson, P., and Ehn, D. C., Itek Internal Report (1975).
8. Nisenson, P., and Stachnik; R. V., Technical Digest from A. A. S. Topical Meeting on Imaging in Astronomy, Cambridge, Mass. (June 1975).

Session 4

PRE-DETECTION COMPENSATION

Session Chairman
R. P. Urtz, Jr.
Griffiss AFB

ACTIVE IMAGE RESTORATION WITH A FLEXIBLE MIRROR*

A. Buffington, F.S. Crawford, R.A. Muller, A.J. Schwemin and R.G. Smits
Lawrence Berkeley Laboratory and Space Sciences Laboratory
University of California
Berkeley, California 94720

Abstract

We have built and tested a 30 cm x 5 cm aperture telescope which uses six moveable mirrors to compensate for atmospherically induced phase distortion. A feedback system adjusts the mirrors in real time to maximize the intensity of light passing through a narrow slit in the image plane. We have achieved essentially diffraction-limited performance when imaging both laser and white-light objects through 250 meters of turbulent atmosphere. The system has yet to achieve its full potential, but has already operated successfully for objects as dim as 5th magnitude. It is presently installed on an equatorial mount at an observatory, and we hope by the time of the conference to present preliminary results with astronomical objects.

Introduction

In a recent article,(1) Muller and Buffington showed that certain image-plane sharpness functions, when evaluated for a system such as a large astronomical telescope, have their absolute maxima only when atmospheric or other phase perturbations have been removed. In addition, through the use of computer simulations, they showed that such sharpness functions providing feedback to a set of discrete moveable optical elements could correct the phase perturbations and yield diffraction-limited telescope images in real time. We have constructed and tested such a telescope system. Our system has achieved essentially diffraction-limited performance from a 30 cm x 5 cm aperture, imaging both white light and laser objects through 250 meters of turbulent atmosphere.

Apparatus

The overall configuration of the telescope is shown in figure 1. The 30 cm diameter f/8 primary mirror is masked at the front of the telescope by an entrance aperture of 30 cm x 5 cm. When corrected, an aperture of this shape should have six times better resolution in one dimension than the other, thus producing essentially a one-dimensional image. Two diagonal mirrors bring the light to a lens which projects an enlarged image onto the sharpness slit with its photomultiplier PM-1. This enlargement, typically 100 times, allows convenient construction and mounting of the various components and slits receiving the light. The six moveable optical elements which comprise the second diagonal mirror are driven by the control electronics. They maximize the signal from PM-1 and thus compensate the atmospheric phase distortions. A pair of steering photomultipliers (PM-3 and PM-4) provide the signal which drives the first diagonal to keep the image centered. The beam splitter directs 20% of the light to another photomultiplier (PM-2) which is used to scan the image in the high-resolution dimension; its output as a function of position is recorded on an X-Y plotter. PM-2 and its slit are mechanically driven across the image by a stepping motor; recording a single image takes typically ten to fifteen minutes, although there is no fundamental limit to much more rapid image recording.

The system uses the sharpness function(1)

$$S = \int dx\, dy\, I(x,y)\, M(x,y) \qquad (1)$$

where x and y denote image plane coordinates, $I(x,y)$ is the image-plane irradiance, and $M(x,y)$ is a mask which is an estimate of the true restored image. A slit in front of PM-1 imposes $M(x,y)$ on the image, so the photomultiplier output yields the measurement of S. The slit width is somewhat less than a full-width-at-half-maximum of a restored, diffraction-limited image. The slit also integrates the light from the direction of poor resolution, thus taking maximum advantage of the asymmetrical resolution of the system. A narrower slit is used in front of PM-2, typically one-half to one-quarter the width of the sharpness-defining slit, so there is negligible contribution to the diffraction pattern from the finite resolution of the scanning data-recording system.

The flexible optics consist of six mirrors driven by piezoelectric "crystals" acting as pistons. Figure 2 shows the construction of this mirror. Six pieces of glass 1.25 cm x 1.9 cm x 0.5 cm were selected for flatness within 1/10 wavelength of light. These were aluminized and placed in a row with 0.15 cm gaps between, in optical contact with a large flat. A ceramic base was prepared which had six mounting columns, each of which could be

* This work was supported by the Energy Resources and Development Agency, and by the National Aeronautics and Space Administration (grant NGR-05-003-553).

pushed by any of three set-screws to adjust the angles between column axes and the rear plane of the base. Six piezoelectric cylinders were glued onto the column ends, and the cylinder ends were lapped into a common plane after the glue had hardened. Finally, the assembly was glued with low expansion epoxy onto the six mirrors that had been placed on the optical flat. The eighteen adjustment screws could then be used, when the epoxy had hardened, to make the planes of the six individual mirrors parallel, and bias voltages on the piezoelectric cylinders could then make the six planes coincident. The piezoelectric cylinders move perpendicular to the plane of the mirrors. Fractional-wavelength mirror movements can be carried out by this system in as little as 0.06 milliseconds, a limit set by the desire to keep mirror operation below the 8.5 kHz resonance of these particular piezoelectric cylinders. Application of ±1000 volts results in a motion of ±2.5 microns.

The first diagonal mirror, used to keep the image centered on the sharpness slit, was mounted on three other piezoelectric columns. These steer the image along the dimension of good resolution by ±10 arc sec with the application of ±1000 volts. This steering capability, although not necessary for the horizontal-path tests to be reported here, is vital to correct steering errors in the equatorial drive of astronomical telescopes. By removing first-order image translation directly, however, the steering does lessen the corrections that the second diagonal must provide, thus slightly improving the resultant diffraction pattern.

Figure 3 shows a block diagram of the electronics. The control circuits introduced a small perturbation (typically 0.05 to 0.10 microns) into the position of each of the six moveable mirrors in sequence; if the "new" sharpness signal was larger than the "old" sharpness signal, the perturbation was left in place--otherwise the mirror was moved back. After one such cycle through the six mirrors, taking typically 4 ms, the sequence was repeated with perturbations in the other direction. A detailed sequence of program steps is given in Table I. Three FET switches select time constants for a Sygnetics 555 integrated circuit which acts as program timer. The 555 steps a 5-bit binary counter which in turn drives two 8 x 32 bit PROMs (Programmable Read-Only Memories). The PROMs contain the program of Table I and also the bits which operate the FET switches and thereby determine the time interval that the program remains on a given step. All PROM outputs are gated to avoid false codes while the 5-bit counter is switching. The "GO TO" instructions of Table I are carried out by a diode network; the network is hard-wired to a plug-in header, and connects to the load inputs of the 5-bit binary counter. The "Mirror Logic" block of figure 3 contains an additional 5-bit binary counter which is advanced or reset by the instructions of Table I. Two additional PROMs driven by this second 5-bit counter contain the mirror movement and replacement sequence. At present, we have operated the mirrors only in a sequential, one-at-a-time mode, but changing to group mirror motion (for example, using Walsh functions) involves only a trivial PROM re-programming and replacement. We have reason to believe that group mirror motion, by increasing the sharpness change caused by a particular perturbation, may enable the system to correct dimmer objects than are currently possible.

Table I. Program Logic for the Flexible Telescope System

Step Number	Instruction
1	Reset Mirror Logic Counter
2	Zero "New" Integrator
3	Integrate Sharpness PM into "New"
4	Load "New" into "Old"
5	Advance Mirror Logic Counter
6	Move Mirror
7	Zero "New" Integrator
8	Integrate Sharpness PM into "New"
9	If "New" less then "Old" go to Step 13
10	Load "New" into "Old"
11	Advance Mirror Logic Counter
12	Go to Step 15
13	Advance Mirror Logic Counter
14	Replace Mirror
15	If Mirror Logic Counter not finished, go to Step 5
16	Go to Step 1

The "Move Mirror" commands (steps 6 and 14) gate the proper polarity constant-current sources, which feed into storage capacitors for each of the six mirrors. The voltages of the integrated result stored in the capacitors are then amplified 1000 times by six DC-coupled transistor-tube amplifiers for placement on the six piezoelectric drivers. The tubes are 6BQ6's with a 2800 volt supply. The mirror positions are thus proportional to the voltages on the capacitors, and the gains of the individual channels are set so a given mirror perturbation instruction moves any one of the mirrors by the same amount. The size of a mirror perturbation is determined by the length of time spent upon the

"Move Mirror" instruction. This time is selected by the FET switches for the 555 timer, and can be set anywhere between 60 and 600 μ sec, corresponding to mirror motions from 0.025 to 0.25 microns.

The "Integrate Sharpness" instruction of Table I is similarly variable from 10 to 4000 μ sec; for all of the runs presented here it happened to be set at 440 μ sec. All of the other steps of Table I take a fixed 10 μ sec.

In figure 3, the "Comparitor" block passes the sharpness measurement from PM-1 through a FET switch which is open only during the "Integrate Sharpness" instruction from the Program Logic block. When the gate is open, the signal is integrated by the "New Light" capacitor. A second FET switch transfers the voltage on the "New Light" capacitor to an "Old Light" capacitor, and a third FET switch dumps the charge on the "New Light" capacitor. The commands for actuating all of these FET switches come from the Program Logic, which executes instructions according to Table I. A LM311N comparitor monitors both "New" and "Old" capacitor voltages to determine whether "New" has exceeded "Old". If during one of the integrations it has, a latch is set which (in step 9) causes the program to advance to the next mirror without replacing the one which was just moved in order to make the improvement.

A variant on the instruction sequence of Table I was tried which allowed repeated iteration on a single mirror until the sharpness no longer increased. No difference in performance of the system was detected when this variant was in operation.

System Alignment and Calibration

The system was aligned by using unresolvable light sources (laser or white light) placed 250 meters away from the laboratory window. The mechanical screw adjustments on each of the six moveable mirrors allowed us to tilt each of the mirrors until their Airy disks overlapped. Coincident Airy disks guarantee that the planes of the individual mirrors are parallel, or at least parallel to a spherical surface that can be corrected for by a slight focus adjustment with the projection lens. Because the mirrors had been pre-aligned by being pressed against an optical flat, each mirror typically needed only a small screw adjustment of the order of the size of its Airy disk. To insure that the planes of the individual mirrors are coincident, and not just parallel, they were covered by a mask having a narrow slit across the middle of each of two chosen adjacent mirrors. The high-voltage bias setting of one of these mirrors was then adjusted to shift the white light fringe over to the center of the image as marked by the sharpness slit. This procedure was repeated for each pair of adjacent mirrors; the adjustments needed were typically a wavelength of light.

To calibrate the angular scale of the recorded data, a "comb" was placed over the aperture which blocked out all but the middle third of each of the six moveable mirrors. The comb converted the instrument into a six-slit interferometer with accurately known slit spacing. This spacing, together with the known laser wavelength (0.6328 microns) determines the angular distance between adjacent peaks on the interference pattern. Figure 4 shows the result of a typical calibration run, as recorded by sweeping PM-2 over the image, and with PM-1 providing image stabilization.

The laser was also used to determine the size of measured sharpness change for full aperture resulting from various mirror perturbations. Table II presents the observed ΔS/S for a corrected image, with the indicated perturbation sizes. The observations were made by simply monitoring the PM-1 output with an oscilloscope while the feedback was on (on a relatively calm day) and observing the sharpness decrease accompanying the unsuccessful mirror perturbations. The results of Table II are about as we expect from computer simulations.

Table II. Changes in Sharpness S Resulting from Mirror Perturbations

Mirror Motion (Microns)	Fractional Change in Sharpness ΔS/S
0.08	0.15 ± 0.10
0.10	0.25 ± 0.10
0.14	0.40 ± 0.10
0.24	0.65 ± 0.10

Horizontal-Path Test Results

The light sources used in alignment and calibration provided a good preliminary test of the ability of the telescope system to compensate for atmospheric turbulence. The light from these travelled along a horizontal path which bridged a canyon. This test range had a variety of seeing conditions which depended on the weather, and provided a good simulations of observatory conditions except for scintillation.

Figure 5 shows the image of the laser as recorded by the scanning photomultiplier, both with and without the image-sharpening feedback. Comparison with a Monte Carlo calculation (dotted line) shows that the system is performing as expected. The small bumps near ±2.5 arc sec are the innermost of a series of secondary maxima (several are visible in the figure) caused by the regularly spaced gaps between the six moveable mirrors.

To protect the photomultipliers from the full laser intensity, we placed neutral-density filters in the optical path and diverted some of the light with beam splitters. Typical attenuation was 2×10^4. To determine the least correctable brightness of object, we introduced additional filters and observed the resulting image degradation. Using the visible bandpass relation between astronomical magnitude m and brightness B,

$$B = (4 \times 10^6) 10^{-m/2.5} \text{ photons/cm}^2\text{sec} \quad (2)$$

the laser corresponded to a magnitude -12 object. The laser image showed stabilization and good image quality until the light had been attenuated by a total factor of 5×10^6. This light level corresponds to about 100 photoelectrons in PM-1 for each integration time of 0.44 ms, which is close to the predicted 36 photoelectrons for a six-mirror system given by the simple analysis in ref. 1. At these low light levels, significant noise was contributed by photomultiplier dark-current, as well as photoelectron fluctuations. Figure 6 shows the deterioration of the image at low light levels, as a function of the numbers of photoelectrons produced by the image during one integration time.

To get a quantitative measure of the time structure of the seeing, we turned off the feedback and analyzed the PM-1 output I(t) with a Hewlett-Packard correlator Model 3721A to determine the integral

$$A(\tau) = \int_0^T dt\, I(t) \cdot I(t+\tau) \quad (3)$$

where $T >> \tau$ and A(0) is normalized to unity. Figure 7 presents typical measurements of $A(\tau)$ with various seeing conditions. Image restoration was not generally possible when the characteristic coherence time of the seeing (defined to be that value of τ for which $A(\tau) = 0.5$) was shorter than the time for three passes through the mirrors. For the results presented here a single pass through the mirrors took about 4 msec. With a window isolating the laboratory room from the outside, the coherence time was longer than 12 msec (and hence the image correctable) about half the time.

We also tested the image-sharpening system on a white-light source. An incandescent bulb with a filament one millimeter in diameter was masked to 1/8 mm in apparent size in order to test fully the resolving capability of the telescope. Figure 8a shows the uncorrected image, and 8b the image with the feedback on. As with the laser, the system yielded an essentially diffraction-limited image.

Observatory Tests

The telescope system has recently been moved to the Leuschner Observatory, where it has been installed "Piggy-back" onto the 20-inch telescope. Thus provided with an equatorial drive, we hope by the time of the conference to present stabilized images of some of the brightest stars in the sky. The observatory location (a few miles to the east of Berkeley in low rolling hills) is not noted for good seeing, and we have already found that coherence times of 4 to 6 msec and seeing disks of 4 arc sec are common in January. We have, however, already observed significant image sharpening with the star Sirius.

References

1. R.A. Muller and A. Buffington, J. Opt. Soc. Am. 64, 1200 (1974).

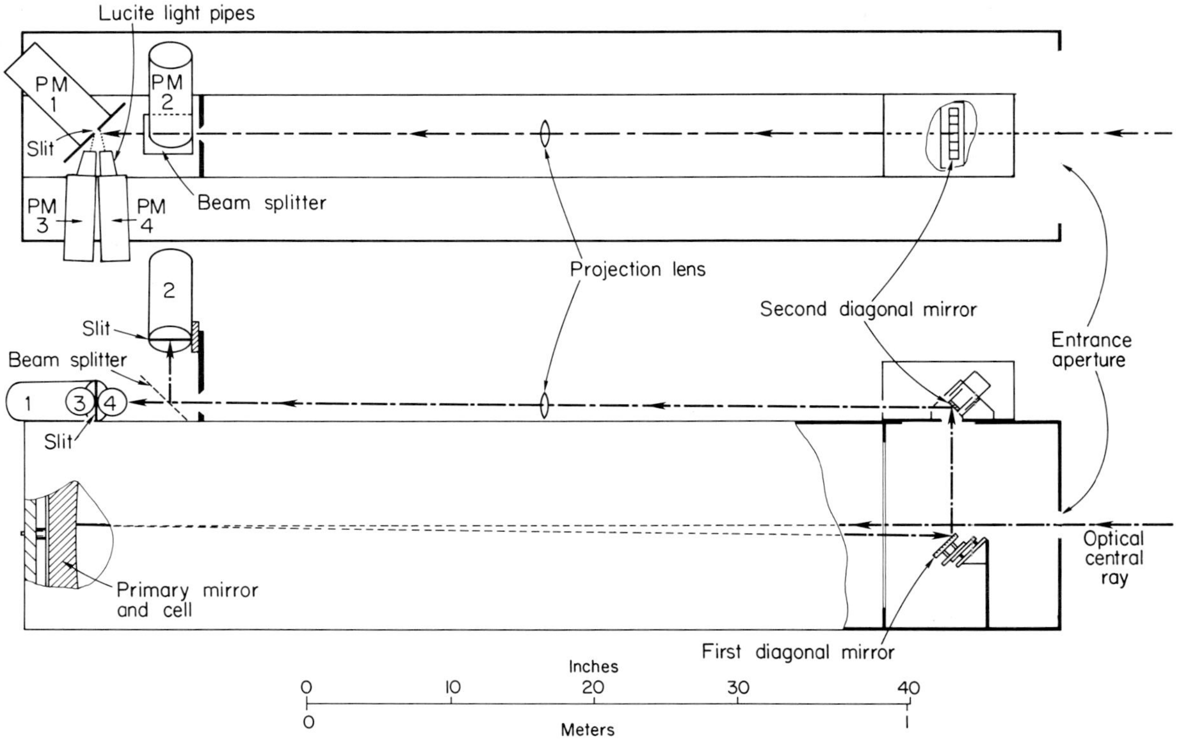

Figure 1. Mechanical layout of the apparatus.

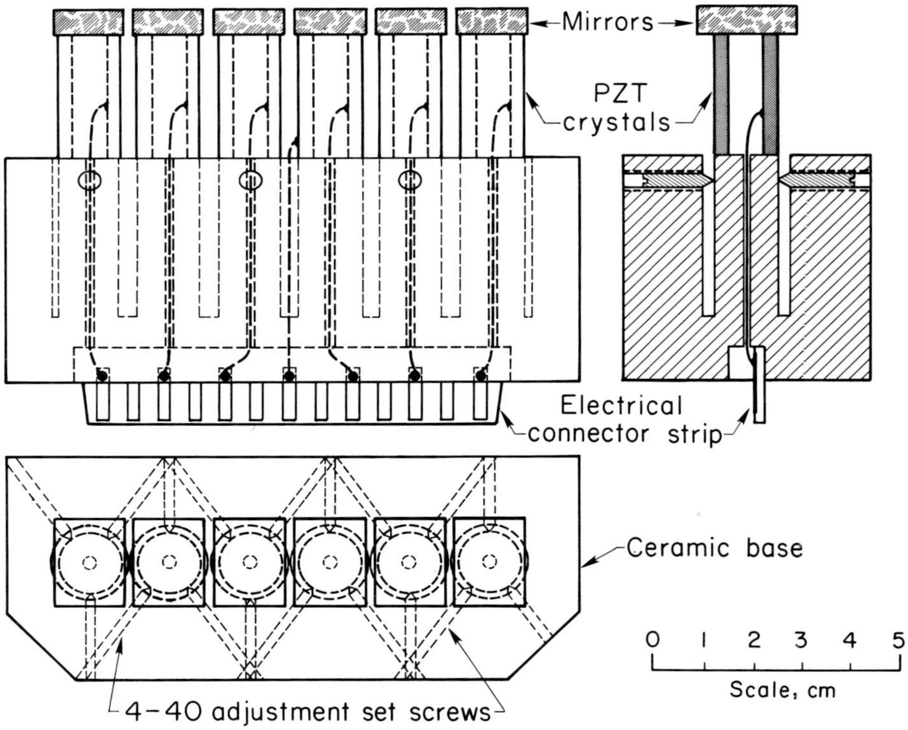

Figure 2. Design for the flexible mirror. Mirror plane angles are adjusted by the set-screws, while positions are adjusted by bias voltages placed on the piezoelectric crystals. Angles and mean mirror positions are fixed in the initial calibrations of the mirror, while rapid perturbations in mirror positions about the mean perform the active wavefront phase correction.

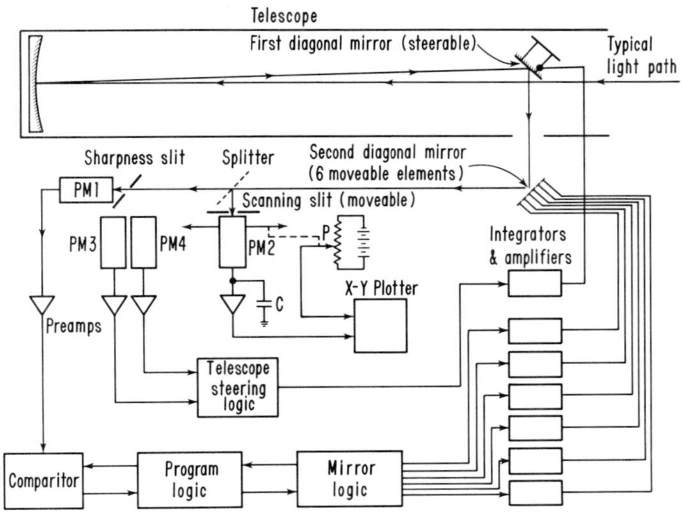

Figure 3. Block diagram of the electronics

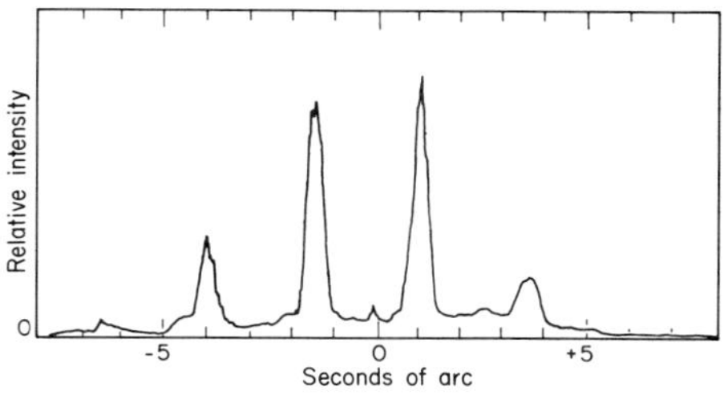

Figure 4. Interference fringes resulting when the system was operated as a six-slit interferometer: feedback on, and system illuminated by the laser. With a 5 cm spacing between the slits, and the laser wavelength of 0.6328 microns, the interference maxima are expected to be separated by 2.6 arc sec.

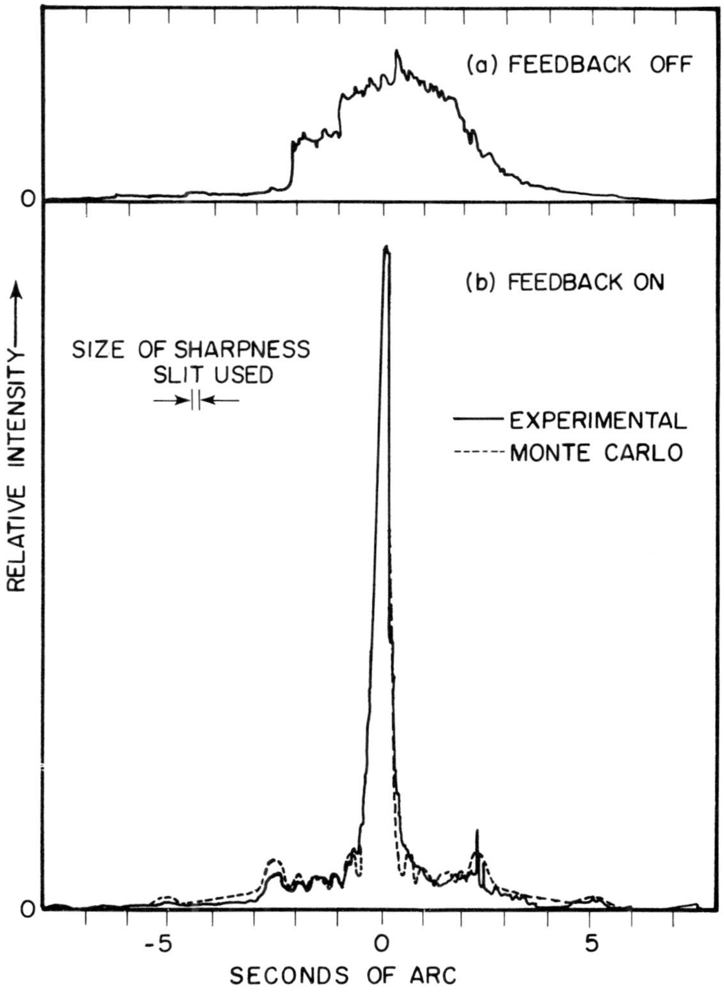

Figure 5. Images of laser light viewed through 250 m of turbulent atmosphere. The agreement between corrected image and Monte Carlo calculation is good, and the corrected central diffraction peak is nearly a decade improved over the uncorrected image.

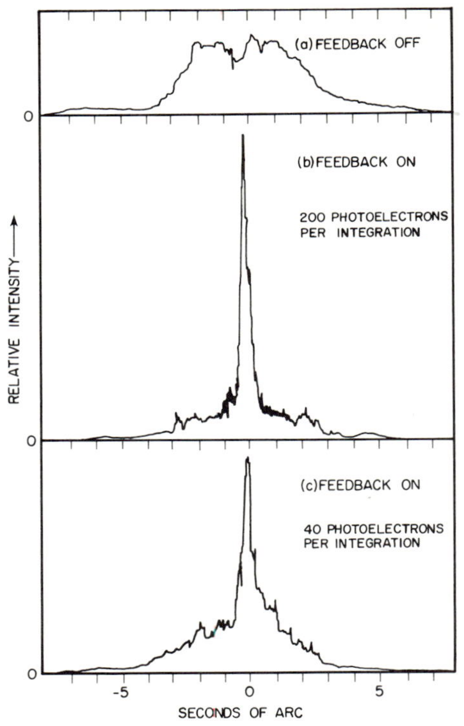

Figure 6. Determination of dimmest correctable objects. The conditions are similar to those of figure 5, except neutral-density filters have been placed in the optical path to reduce the amount of light. The image quality in (b) is nearly as good as that of figure 5 even though the amount of light has been reduced by 100 times. However, the image quality has degraded when the light was further reduced in (c).

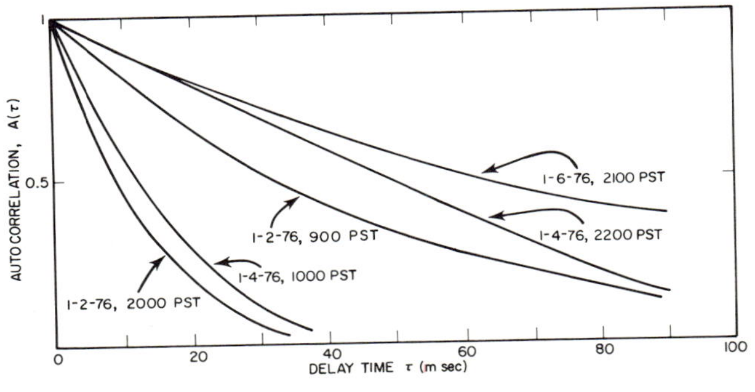

Figure 7. Measurements of the autocorrelation function $A(\tau)$. These curves show the characteristic coherence time for speckles from the uncorrected laser image drifting across the sharpness slit. The curves chosen represent a typical range of seeing experienced at our laboratory. These curves illustrate the wide range of conditions that can occur within a relatively short time period.

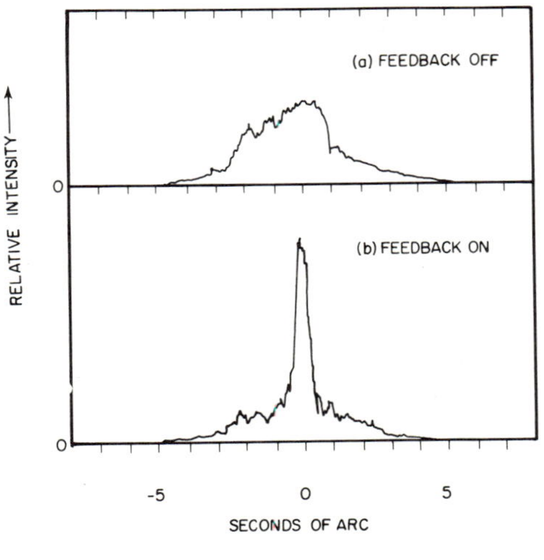

Figure 8. White light images. An incandescent bulb with a 1 mm filament was placed 250 m from the telescope. A narrow mask was placed in front of the bulb to make an unresolvable object.

USING MEMBRANE MIRRORS IN ADAPTIVE OPTICS

Martin Yellin
The Perkin-Elmer Corporation
Norwalk, Connecticut 06856

Abstract

This paper describes the utilization of very thin, electrostatically deflected membranes as an active optic in an image compensation system. The key design considerations are given in terms of deflections, frequency response, and drive signals. The advantages of a membrane are given in terms of its transfer characteristics, low voltages, zero hysteresis and its ability to accommodate hundreds of actuators. Pertinent performance data is presented. This paper also discusses the manufacturing techniques that are used to generate membrane mirrors of different materials, thicknesses, and geometries. The manufacturing technique is relatively simple and inexpensive and leads to a rugged active optic that is ideal for use in an image compensation system, where the correction of many waves is required.

What is a Membrane Mirror

The basic implementation of a membrane mirror is shown in Fig. 1. The membrane, typically 0.5 μm to 1 μm thick, is positioned between a single transparent electrode containing a bias voltage (V_o) and, on the other side, a region of actuators consisting of conducting pads having a voltage $V_o \pm V_{signal}$. The membrane is at ground, and the spacing between the membrane and the electrodes is approximately 25 μm. With no signal V_s, the membrane experiences zero net force, and there is no deflection. The membrane will stay flat within $\lambda/20$ rms. When a voltage (V_s) is applied to any electrode, there will be a local deflection, typically of $\pm$0.5 wave with $\pm$80 V applied. By driving many actuators, a total deflection of many waves is readily accommodated.

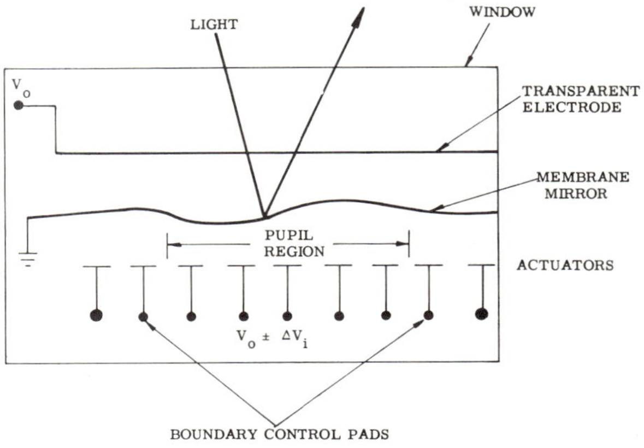

Fig. 1. Membrane Actuator Drive Schematic

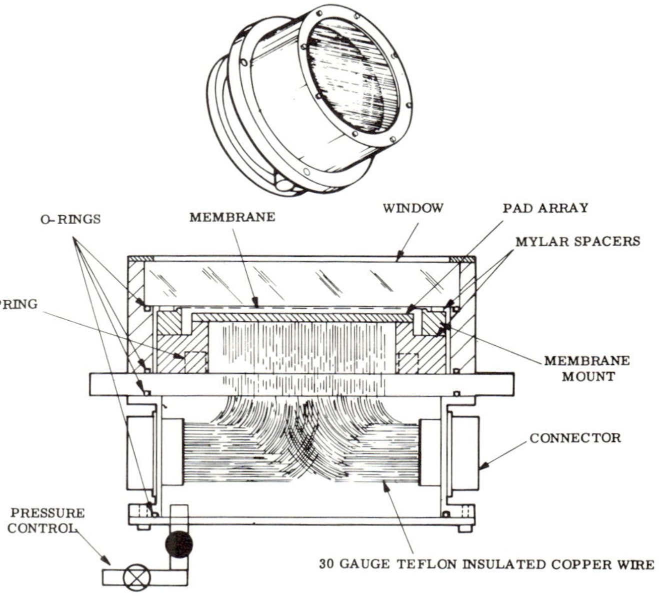

Fig. 2. Wavefront Corrector

Fig. 2 is a schematic of a 100 mm membrane assembly. The interior of the membrane assembly is evacuated to about 20 mm Hg, which has been established by the damping requirements for the assembly. The membrane electrodes are generated as a printed circuit that is polished flat to within 0.5 μm. One of the key features of the membrane is the window, which is an AS-3 conductive transparent coating. Fig. 3 is a photograph of a typical membrane assembly.

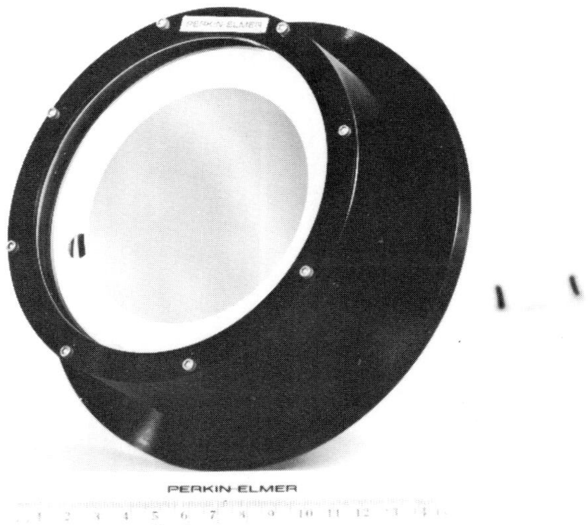

Fig. 3. Adaptive Mirror Assembly

As will be shown in the next section, the static displacement of a membrane is related to the force density or voltage by the relationship: $\nabla^2\nu$ = force density, where ν = displacement and ∇^2 is the familiar Laplacian function. By comparison, a solid mirror bears the relationship: $\nabla^4\nu$ = force density. The Laplacian relationship of a membrane greatly reduces the signal processing required to generate reconstructed wavefronts from incoming wavefront measurements. Another advantage of a Laplacian system is the spatial averaging it performs, effectively filtering and inherently minimizing mean squared error.

Due to the membrane's low mass, there is no hysteresis; low deflection voltages are required; and high resonant frequencies can be achieved (> 5 kHz). Finally, the fabrication of such assemblies is simple and inexpensive, leading to very reliable and rugged mirrors relatively impervious to their environment.

Key Membrane Equations

Membrane Transfer Function

The basic membrane differential equation is given by:

$$\frac{\partial^2 \nu}{\partial t^2} = \frac{T}{D} \nabla^2 \nu \tag{1}$$

where

ν = vertical displacement
t = time
T = tension/length
D = mass/area

With an external stress (F) applied to the membrane, the equation of motion is

$$\frac{\partial^2 \nu}{\partial t^2} = \frac{T}{D} \nabla^2 \nu + \frac{F(r,t)}{D} \tag{2}$$

If we consider simple harmonic time dependence such that $\nu(r, t) = \nu(r)\, e^{-jwt}$ and $F(r, t) = F(r)\, e^{-jwt}$, then

$$\frac{\partial^2 \nu}{\partial t^2} = -\omega^2 \nu(r) = \frac{T}{D} \nabla^2 \nu + \frac{F(r)}{D} \tag{3}$$

or

$$\nabla^2 \nu + \frac{D\omega^2}{T} \nu = -\frac{F(r)}{T} \tag{4}$$

which is the scalar Helmholtz equation.

If F(r) is a point source, the solution is the membrane impulse response function (Green function). For an infinite boundary membrane, the Green function is:

$$g(r-r', \omega) = j\pi H_0^{(1)} \left(\sqrt{D/T}\,\omega[r-r']\right) \tag{5}$$

where H_0^1 is the zero order Hankel function of the first kind. If one takes the spatial Fourier transform,

$$G(k, \omega) = \frac{4\pi}{k^2 - \frac{D}{T}\omega^2} \tag{6}$$

where k = spatial frequency and ω = temporal frequency.

These equations neglected damping. Viscous damping to any desired damping ratio is controlled by varying the air pressure in the cavity (typically 20 mm Hg for critical damping).

Note that in the steady state:

$$\frac{T}{D} \nabla^2 \nu = \frac{-F(r)}{D} \tag{7}$$

or

$$\nabla^2 \nu = -\frac{F(r)}{T} \tag{8}$$

and

$$\frac{F(r)}{T} = \frac{\epsilon_o V_s V_B}{\ell^2} \tag{9}$$

for an electrostatic deflector, where $\epsilon_o = 8.85 \times 10^{-12}$ f/m, ℓ = separation between electrodes, V_s = signal voltage across electrodes, and V_B = bias voltage across electrodes.

Thus, for the reference design, if V_s is made proportional to the Laplacian of the measured wavefront, the deflection (ν) would equal the desired reconstructed wavefront. In implementation in an image compensation system, a Hartmann wavefront sensor can be used to obtain local slopes (∇W) of a wavefront. By taking the spatial divergence of the slope measurements ($\nabla \cdot \nabla W = \nabla^2 W$) and setting that signal proportional to V_s, one can achieve proper wavefront reconstruction.

Deflection Equation

$$\sigma = \frac{\epsilon_o V_s V_B r^2}{4T\ell^2} \tag{10}$$

where

σ = local deflection over a pad (meters)
ϵ_o = 8.85×10^{-12} farads/meter
r = electrode radius (meters)
T = linear tension (Newtons/meter)
ℓ = gap (meters)

Typically, $\sigma = 0.3\,\mu$m for $V_s = 80$ V, $V_B = 200$ V, r = 1.2 mm.

Frequency Response

$$f_o = \frac{2.4}{\pi D}\sqrt{\frac{T}{\sigma}}$$

where

D = membrane diameter (m)
T = linear tension N/M = stress times thickness
σ = mass/unit area kg/m^2 = density times thickness
f_o = first resonant frequency of the circular membrane

Material Considerations

Perkin-Elmer has successfully manufactured membranes as thin as 0.25 μm with diameters of up to 125 mm. The key considerations for material selection are:

(1) Manufacturability

(2) Desired frequency response, which is related to high yield strength and low density

(3) Ability to be manufactured optically flat

(4) Ability to be coated with highly reflective coatings

Membranes have been manufactured at Perkin-Elmer using titanium, titanium alloys, nickel, beryllium, molybdenum, and silicon. For a high frequency mirror, three membrane materials have been used, as shown in the following table.

Material	Thickness	Frequency for 75 mm Membrane
Titanium Alloy	0.25 → 1 μm	> 5 kHz
Titanium	0.25 → 1 μm	> 4 kHz
Nickel	1 μm → 20 μm	2 kHz

Membrane Production Procedure

Most membranes are produced by evaporating the material onto a substrate in a vacuum chamber; this is shown in Fig. 4. The exact procedure consists of the following steps:

(1) Polish substrate (glass or CaF_2)

(2) Install on gallium-wetted block and pump down

(3) De-gas evaporation sources

(4) Deposit CaF_2 parting layer (if glass substrate)

(5) Heat substrate to 500°C

(6) Set titanium evaporation rate (and rod feed rate if alloy)

(7) Open shutter until film reaches desired thickness

(8) Shut down and allow to cool

(9) Remove substrate from chamber

(10) Examine surface by Nomarski microscopy; measure film thickness

(11) Float film off of substrate

(12) Mount and tension film

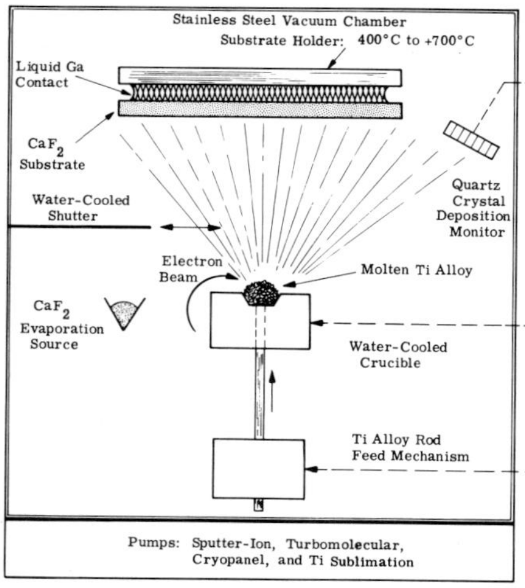

Fig. 4. Titanium Membrane Facility

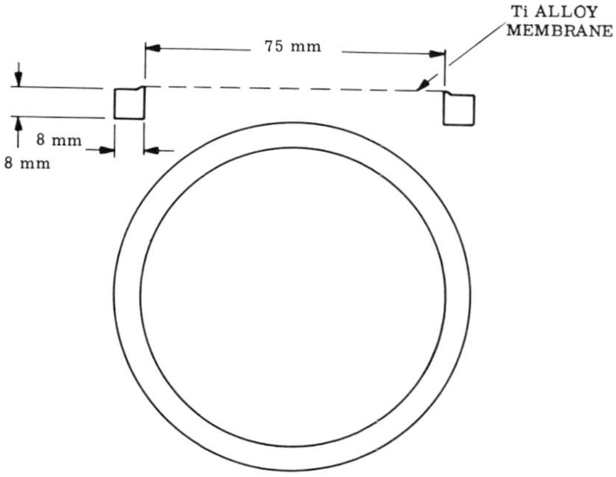

Fig. 5. Membrane Ring
(Stainless Steel, λ/25 Polished Ring)

Once a membrane is made, it is tensioned onto a ring and epoxied in place as shown in Fig. 5. The tensioned ring is then located over the printed circuit electrodes using a special precision separator. It takes about one hour to generate a membrane and about two days to assemble an entire unit.

Membrane Performance

Figures 6 through 9 show some of the performance characteristics of a particular membrane, namely a 75 mm titanium alloy membrane. Fig. 6 shows damping characteristics versus viscosity. The viscosity value corresponding to a damping ratio of 0.72 occurs at a pressure of 20 mm Hg. Fig. 7 shows deflection growth versus time and shows that the membrane exhibits the -log r^2 profile desirable for spatial filtering. Fig. 8 shows membrane error versus time. Fig. 9 shows the extreme case of commanding the membrane to tilt back and forth every 100 μs full deflection.

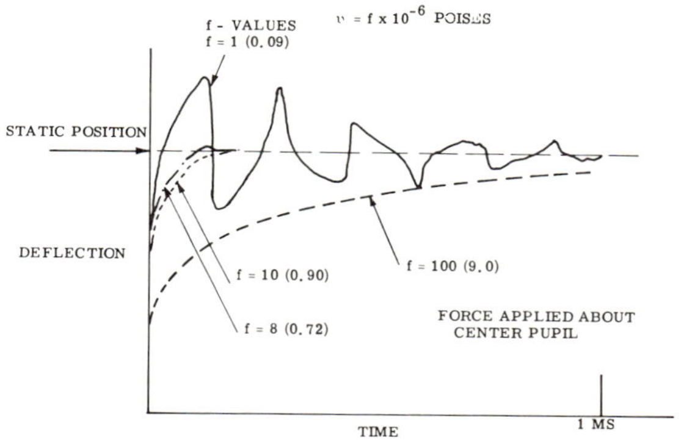

Fig. 6. Deflection vs Time at Center Pupil with Viscosity Variation, Damping Ratio is in Parentheses Adjacent to f-value: f(γ)

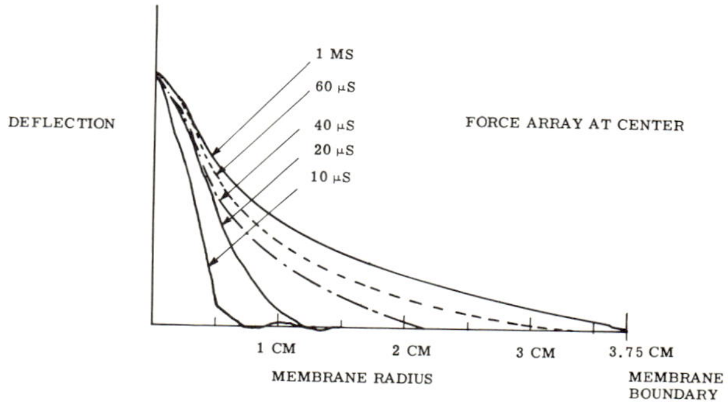

Fig. 7. Deflection Growth vs Time

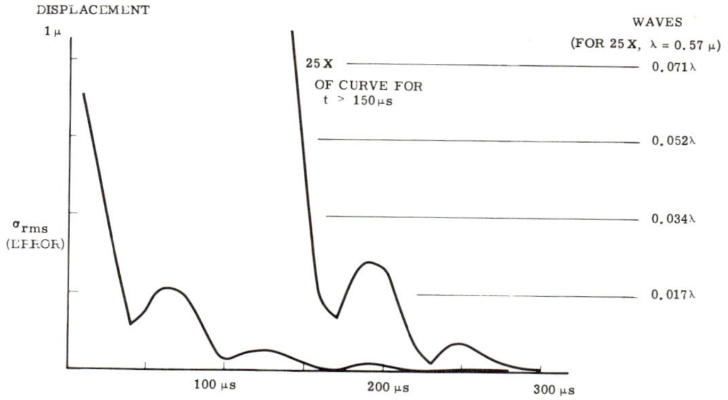

Fig. 8. σ_{rms} vs Time for Membrane Changing from Zero Force to Tilted Plane State (σ_{rms} Measured Relative to Static Position)

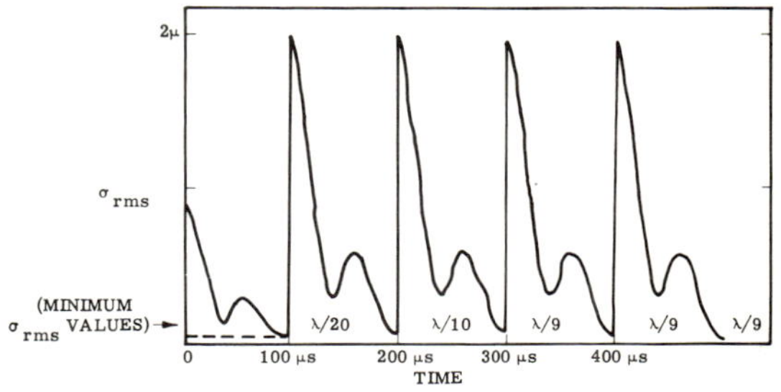

Fig. 9. σ_{rms} vs Time for Membrane Changing from Positive to Negative Tilt States Every 100μ s (σ_{rms} Measured Relative to Static Position)

Use of Membrane Mirrors in an Image Compensation System

Figure 10 shows how a membrane mirror can be used to correct aberrations in an image compensation system.

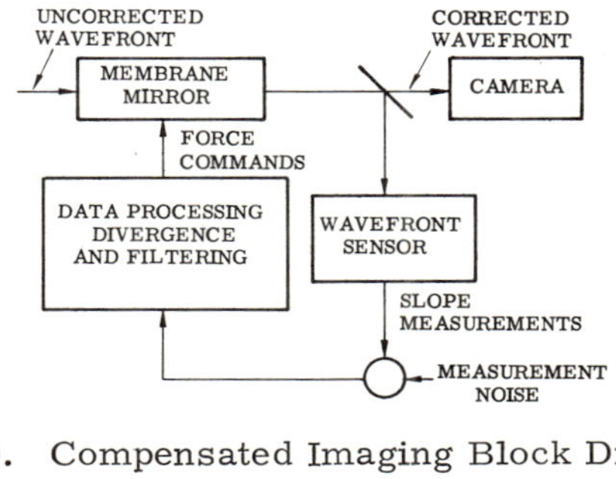

Fig. 10. Compensated Imaging Block Diagram

Conclusions

Membrane mirrors, both in their physics and in their engineering, lend themselves optimally to image compensation systems, especially when confronted with spatially correlated wavefront disturbance. They are relatively inexpensive and highly reliable, and their low voltage requirements lend themselves to a wide range of applications, especially in space environments.

WIDEBAND ADAPTIVE OPTICS FOR IMAGING*

Dr. Julius Feinleib and J. W. Hardy
Optical Systems Division, ITEK Corporation
10 Maguire Road
Lexington, Massachusetts 02173

Abstract

To obtain high resolution images from large aperture optical systems, it is well known that the optical path must be corrected before the image is recorded and the phase information essentially lost. This pre-detection compensation requires fast response, adaptive optical elements in order to correct for optical path errors introduced by a turbulent atmospheric path. Fundamental concepts and operation of a laboratory system for wide band real time atmospheric compensation is discussed. The RTAC system operates closed-loop and automatically locks-in to correct phase aberrations of a rapidly varying turbulent atmosphere.

Introduction

It was recognized very early in the history of optical observation that the resolution of distant objects was limited by the low light fluxes received from the distant source and by the imperfections in the optical path.

Increasing the aperture of the optical system was known to increase both the light gathering power and the theoretical resolution of the observing system and improving the quality of the mirrors and lenses were the primary means of extending the system resolution to the theoretical limits. As the apertures became larger than 20 cm, system capabilities did not improve proportionally because of bad "seeing".

Even Newton recognized that for a ground-based observer, little could be done to improve the optical path imperfections caused by turbulence in the line of sight through the atmosphere. He suggested that large telescopes be placed on high mountain peaks to avoid as much of the atmospheric path as possible. This thinking has led to the present concept of a large space telescope where observations may be made beyond atmospheric interference. Several alternatives to avoiding the atmosphere have been proposed.

Attempts have been made to use large digital computers to process images that have been recorded with ground-based optical systems. So far results have been unimpressive, largely because it is difficult to deconvolve the image information from the random phase aberrations in the optical path, and present computers and algorithms have limited processing capability to keep up with the information rates of continuous observations.

The fundamental problem with these processing systems is that once the wavefronts arriving from the object are recorded, the phase information is lost because the recording medium in normal optical systems only records the intensity average of the complex wavefront. Path length aberrations affect the phase portion of the wavefront and the information needed to restore the true wavefront is essentially lost.

Optical designers were well aware of this problem. Large aperture optical systems are sophisticated and expensive because of the attempt to design out or compensate for all phase aberrations in the optical path and maintain phase coherence throughout the system. Phase compensation plates are useful only for the static aberrations, but the principle is fundamental: the path length errors must be compensated before the image is recorded and the phase information lost. For the atmospheric problem where the distortions are occurring randomly and at a high rate, the usual solution of polishing special phase plates could not be used and the pre-detection compensation principle required the evolution of a high speed adaptive optics technology.

* This work sponsored by the Air Force Systems Command's Rome Air Development Center, Griffiss AFB, NY and is sponsored by DARPA.

Wide Band Pre-Detection Compensation

Wide band pre-detection compensation required the development of three components: a high speed wavefront sensor to detect, in real-time, the phase or path errors caused by the atmosphere or optical system aberrations; a high speed computer to transform this measurement into control outputs; and a high speed deformable optical element which is controlled to perform the optical path compensations in real-time before the image is recorded.

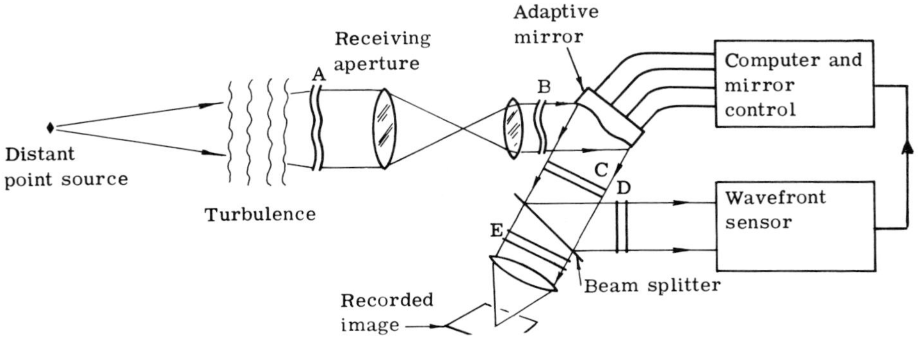

Fig. 1 — Adaptive optics system

A schematic of a general pre-detection compensation system is shown in Figure 1. For illustration, the light from a distant point source is shown passing through the aberrating medium, e.g., the turbulent atmosphere, and is received by the large collecting optical aperture. In addition to the path length aberrations introduced by the atmosphere at A, static path errors are also added by the other optics with the resultant deformed wavefront shown at B. Wavefront B is reflected from the deformable mirror.

In principle, the surface of the deformable, or adaptive mirror is pre-distorted by the control signals from the computer so that it adds and subtracts optical path lengths to each point on the wavefront at B to just compensate the path length errors in B. The result will be the corrected wavefront C which is then recorded.

How does the deformable mirror get the appropriate control information about the wavefront B in time to make corrections to it before C is recorded? The concept is to have the system respond at a high enough frequency so that all the optical path errors, including those arising from the atmospheric turbulence, appear static while the system measures the path errors and applies the correction to the adaptive mirror. With this large bandwidth response, the system will adapt fast enough so that it always keeps up with the changing wavefront at B, and the recorded wavefront C will be optimally corrected except for residual errors in the system.

As shown in Figure 1, the wavefront sensor measures the wavefront after it is reflected from the adaptive mirror. It is thus measuring deviations of the corrected wavefront at D which is a sampled portion of C and the same wavefront E that is recorded. This closed-loop arrangement means that once the system measures the initial correction and applies it to the deformable mirror, it is then sensitive to changes in the wavefront from this initial state and is always applying corrections to lock in on an optimally corrected wavefront. If the loop response is fast enough, then as the path deviations change due to atmospheric turbulence, the adaptive mirror is controlled to keep pace with these deviations.

Whatever wavefront sensor and computer are used in the system depicted, they can only determine at most, the path length or phase deviations from some reference value at each point in the wavefront. It is evident that the wavefront seen by the sensor is disturbed by the unknown effects of the random atmospheric turbulence. How does the system know what the corrected phase should be in order to initialize the system and maintain closed-loop correction? The sensor can measure the total phase at each point but cannot separate out the contribution from the source and the effects of the atmosphere unless some assumptions are made. For the situation depicted in Figure 1, the necessary information about the correct wavefront is readily determined. We have chosen a point source object at a distance large compared to the receiving aperture of the optical system. If there were no atmosphere present, then the point source would radiate a spherical wavefront which arrives at the aperture with only a small curvature - it would be essentially a uniform plane wave across the aperture. The compensation system is designed so that the aperture is imaged onto the adaptive mirror. This image is usually greatly demagnified so that a small adaptive mirror may be used for convenience. Without aberrations, the wavefront incident on the adaptive mirror is also a plane wave. The adaptive mirror will be controlled to a plane surface as there are no deviations from a plane wave to compensate. The wavefront sensor, by whatever means, will measure the wavefront, and with the computer will determine that there is an unaberrated plane wave and so control the mirror to a plane surface. This plane wave is then recorded and gives a point image - the correct image of the distant object.

When there is a turbulent atmosphere between the point source and the aperture, the received wavefront will no longer be a plane wave. First, assume that the turbulence occurs close to the receiving aperture. The path length and, therefore, the phase at each point of the wavefront entering the aperture will deviate from a plane wave condition. The adaptive mirror, being at a conjugate position of the aperture is now required to add or subtract path length to restore the wavefront to a plane wave condition. This is the compensation required to produce a corrected image of the distant point source object. In this case, it is apparent that the wavefront sensor need only measure any deviation from a plane wave in order to compute the correct compensation. A particularly simple computation is required if the wavefront sensor measures the path deviations at points in a plane conjugate to an image of the aperture. Then since there is also an equivalent point on the adaptive mirror for each measured point, control need be applied only at this one point on the mirror to effect compensation - there is a one-to-one mapping of the receiving aperture on the wavefront sensor and on the adaptive mirror. The computer algorithm required to apply the correct control is thus simplified and a very simple analog electronic computer can be used. In addition, if each point on the deformable mirror can be controlled independently of the rest of the mirror surface, then all corrections can be applied to the adaptive mirror directly from each measurement point of the sensed wavefront, allowing parallel addressing of the mirror. Thus, the speed of response of the whole system loop depends only on the speed of response of each measurement-correction channel and is independent of the number of such control channels. This is the key to building a practical real-time correction system with sufficient bandwidth to compensate for atmospheric effects.

We have considered the simple case of a point source and path error occuring near the receiving aperture. If the path errors are distributed over large atmospheric paths, the compensation problem becomes more complex. This problem of isoplanatism has been considered theoretically for many situations and different compensation systems, and it is sufficient to summarize the results here by stating that, in general, as the conditions depart from the simple case considered, the degree to which compensation can be obtained using the simple system depicted in Figure 1, drops gradually until it becomes ineffective. However, for most astronomical situations, where the atmospheric effects are relatively close to the aperture because of the object distances, the plane wave reference condition holds and the simple compensation system can be quite effective if the three major components of the system - the wavefront sensor, adaptive mirror and computer - can be built with sufficient frequency response.

Experimental System

Development of these components and their integration into a system for laboratory demonstration has been accomplished at Itek under sponsorship of the Rome Air Development Center and is described briefly.

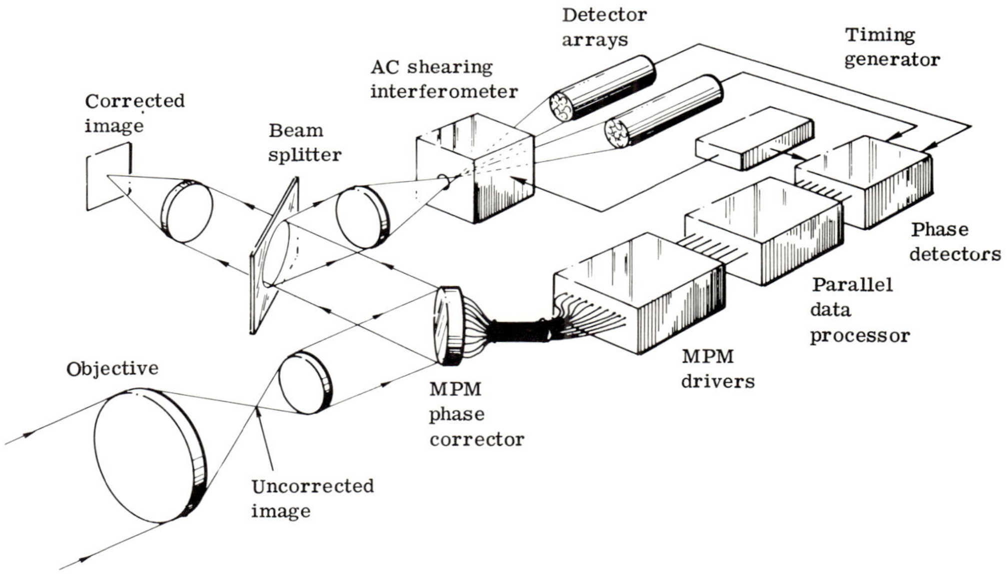

Fig. 2 — RTAC system

Figure 2 shows a block diagram of the Itek RTAC (Real Time Atmospheric Compensation) system.[1] It is basically the system described in Figure 1 using as the system components, a shearing interferometer wavefront sensor, a simple parallel analog computer and a proprietary high speed adaptive mirror, the monolithic piezoelèctric mirror (MPM). These system components where specifically developed to achieve the wide bandwidth required for atmospheric compensation. The laboratory system uses 21 parallel measurement and control channels in a closed-loop system.

Wavefront Sensor

The shearing interferometer wavefront sensor used in the RTAC is basically similar to the shear plates used to measure the surface figure of static optical elements.[2] It was chosen because it is rugged and insensitive to external vibrations - a severe drawback of most wavefront measuring interferometers. The principle of operation is depicted in Figure 3.

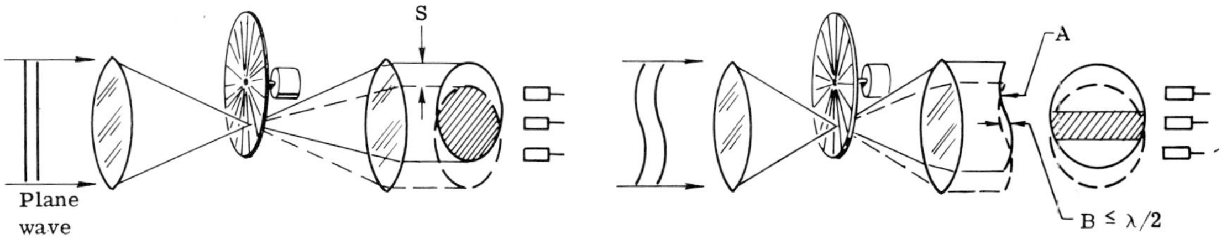

(a) **Plane wave input** (b) **Aberrated wave**

Fig. 3 — Shearing interferometer

The wavefront to be measured is focussed onto a grating as shown. The undiffracted zeroth order light cone emerges and falls on an array of detectors. The first order cone of light is diffracted down and is displaced from the zeroth order by an amount governed by the grating spacing (for illustration, the minus first order is not shown). This diffracted light also falls on the detector array. Since the two light cones arise from the same wavefront and follow nearly identical paths, they arrive at the detector temporally coherent even if the light originated from an incoherent white light source, and therefore, able to interfere in the overlap area. Suppose the wavefront entering the interferometer was a perfect plane wave as shown in Figure 3a. If the grating spacing is such as to cause destructive interference ($\pi/2$ path difference) where light rays from the two cones coincide, then this same interference condition will hold for the whole overlap region and this region will appear as a large dark fringe to the detector array. If the grating is moved vertically by half the grating period, the interference will be constructive and the detector array will see a single bright fringe. Thus, by moving the grating, the light intensity seen by all the detectors will simultaneously vary from bright to dark. The detector outputs will be AC signals and these signals will have the same phase in time if the original wavefront into the interferometer is a plane wave as we have assumed. Thus, the electrical phase of the detector outputs is directly related to the phase differences between points on the incoming wavefront.

Now suppose the incoming wavefront is not a plane wave as shown in Figure 3b. When the slightly shifted light cones are superimposed, there are points along the wavefront where the two interfering light rays have exactly the same path length (A), and interfere constructively, and there are points where the light rays have a half wave path difference (B) and interfere destructively. The fringe pattern on the detector array is shown. While some detectors see intensity maxima, others see intensity minima. If the grating is displaced by half its spacing period, then there is a reversal from constructive to destructive interference in the overlap region of the light cones. The intensity reversal then gives an AC signal on the detectors. As opposed to the case when the light cones were plane waves, the AC output signals of the detectors are no longer in phase with one another and, in fact, the electrical phase of these signals is directly related to the phase differences between two points separated by the shear distance (S) along the vertical direction of the incoming wavefront. The detector output signals can now be processed to obtain the phase deviation from a plane wave at every point on the wavefront where the detectors make measurements.

In the working wavefront sensor, it is necessary to shear the wavefront in both orthogonal x and y directions in order to get a complete two-dimensional wavefront map, and this required two sets of gratings. The value of the shear S required is determined by the magnitude of the distortion in the incoming wavefront. To prevent an ambiguous measurement of phase difference between two adjacent detectors, the phase difference must be kept below 2Π. This can be accomplished by decreasing the shear value until the maximum phase difference between any two adjacent detectors is below 2Π. Gratings with variable spacings are used to accomplish this. The actual hardware implementation of the shearing interferometer and the phase computer have been described in detail previously.[3] Figure 3a above shows the principle actually used to obtain variable shear in the laboratory demonstration system. The system uses a Ronchi radial grating which is rotated by a small motor. The light is focussed on the grating, and the shear value is varied by moving the grating assembly in the direction along the radius of the wheel, without moving the focussed spot. The grating spacing seen by this focussed spot is thus variable from the minimum value near the circumference of the wheel to the maximum near the wheel center. Because the grating lines are closely spaced and the spot small, the non-parallelism of the grating lines presents no problem. The RTAC system using these radial gratings was capable of computing phase deviations at each of 21 wavefront points

Adaptive Mirror

In order for the system to operate with sufficient bandwidth for atmospheric compensation, an adaptive mirror is required which is capable of applying path length compensations at each position in the aperture where wavefront measurements have been made and with a response compatible with the measurement loop response. Adaptive mirrors have been built with electrical control, but they have not been suitable for atmospheric corrections for several reasons. Previous deformable mirrors with continuous reflecting surfaces did not have the frequency response for real-time correction; they used individual actuators at each control position on the mirror, and while these produced large deflections, it was technically difficult and expensive to maintain the optical tolerances with an array large enough to be useful for practical optical systems. Segmented mirrors have been built which achieve the frequency response and required deflection at each point. They have the advantage over previous continuous deformable mirrors in that each segment is separately addressed and acts independently of all the other mirror elements. This allows the wavefront sensor and computer to address each point in parallel and independently without going through time consuming processing if the mirror elements were to interact. The major disadvantage of the segmented mirror is that because the elements are entirely independent, it is possible to have the individual elements displaced so that there are discontinuous steps in the mirror surface. Most wavefront sensors are not capable of measuring phase differences between elements of more than 2π so that without complicated additional control loops, the segmented mirror becomes impractical to implement. What these considerations lead to is that the adaptive mirror must meet stringent requirements. 1) It must have a continuous mirror surface; 2) there must be a sufficient number of separately controllable elements of the mirror so that the expected spatial frequency deviations across the receiving aperture caused by the atmosphere can be compensated with an equivalent spatial frequency configuration on the adaptive mirror: 3) the amplitude of deflection must be at least one wavelength difference between elements and a total deflection amplitude across the mirror of several wavelengths of the light used in imaging is desirable; 4) it must have a frequency response compatible with the turbulence bandwidth and measurement loop; 5) the elements must be essentially independently controllable, but there must be enough interaction so that the mirror takes on smooth figures without sharp discontinuities; 6) the mirror must be stable over a long enough time period to be useful in practical compensation systems. Looking at these severe requirements which must be met simultaneously by a single device, it is not surprising that no atmospheric compensation system was demonstrated before the RTAC system and before the monolithic piezoelectric mirror (MPM) was developed.

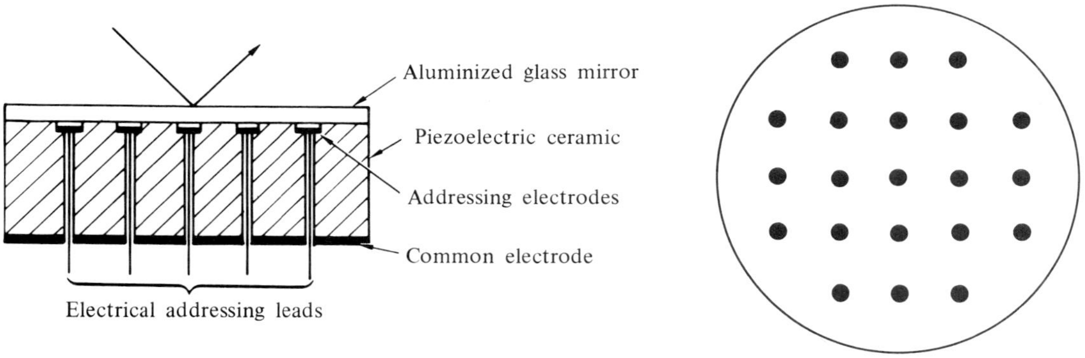

Fig. 4 — Monolithic piezoelectric mirror

The MPM is shown schematically in Figure 4. It is fabricated from a single solid disc of PZT material which has been polarized through its thickness. The active electrodes are brought to the mirror surface through holes bored through the disc, and a ground electrode is connected to the entire bottom surface of the disc. The continuous mirror surface consists of a thin, aluminized glass mirror which has been cemented to the top surface of the PZT disc. With this monolithic construction, it is relatively easy to increase the number and density of addressable electrodes by simply boring more lead holes into the disc.

Although, PZT has been used for a long time as a mirror actuator device, the principle of operation used in the MPM is unique.[5] It was found that the surface of a polarized piezoelectric disc will deform in direct proportion to the voltage applied at each point on the surface when the back surface is held at a uniform ground potential. Thus, it is possible to control the figure of the top surface of the PZT block (and the thin mirror attached rigidly to it) while the bottom surface remains flat. The piezo-electric forces are distributed so that no local forces appear on the bottom surface. The bottom can be left free with no substrate support since no forces occur on this surface. The top surface spatial resolution was found to be smaller than 0.1 mm, so that electrodes could be closely spaced on the top surface, and by applying different voltages to these electrodes, nearly independent deflection of the area around the electrodes are achieved. Examples of such deformations are shown in Figure 5. The electrode

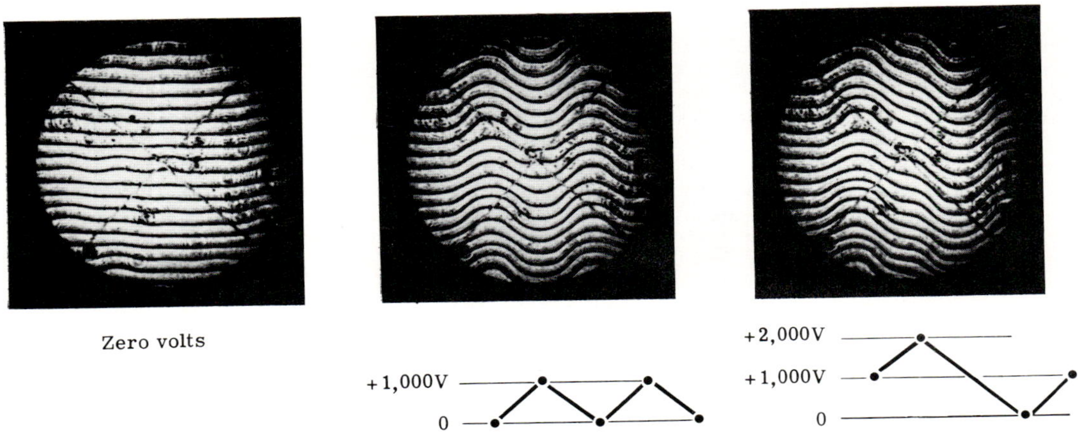

Fig. 5 — 21-element MPM

configurations in the MPM are designed so that the influence function or surface motion overlap from one electrode region to the next gives a smooth surface when adjacent electrodes are addressed. In the laboratory RTAC system, 21 electrodes were placed in a 1-inch diameter MPM. This mirror has operated at frequencies beyond 30 kHz and has deflection amplitudes of at least 2 waves between adjacent electrodes. The device is small, rugged and stable, and is a practical wideband adaptive mirror required for atmospheric pre-detection compensation.

The RTAC system has operated in the laboratory with both static path distortions and hot plate simulated turbulence. In operation, it was found that the shearing interferometer and the closed-loop performance was stable so that the system would automatically initialize the closed-loop correction, lock-in, and follow a rapidly varying turbulence distortion. When the phase aberration exceeded the dynamic range of the system, the system broke lock, but immediately and automatically acquired lock within a few cycles of the measurement loop. Work is now continuing to achieve higher performance levels with a larger number of compensation channels.

The authors wish to acknowledge the contributions toward implementation and the continued progress in the system engineering by C. Koliopoulos, J. Lefebvre, P. Cone, N. Albertinetti and S. Moody. We are grateful for the support of the Rome Air Development Center, and, in particular, the contributions of R. Urtz and D. Hanson.

References

1. Hardy, J.W.; Feinleib, J and Wyant, J.C., "Real Time Phase Correction of Optical Imaging Systems" Proceedings OSA Meeting on Optical Propagation through Turbulence, Boulder, Colo., (July, 1974).

2. Wyant, J.C., "White Light Extended Source Shearing Interferometer", Applied Optics 13, 200 (Jan., 1974).

3. Hardy, J.W., "Real-Time Optical Wavefront Correction", IEEE National Telecommunications Conf. (NTC), New Orleans, La. (Dec., 1975).

4. Feinleib, J.; Lipson, S.G. and Cone, P.F., " Monolithic Piezoelectric Mirror for Wavefront Correction", Applied Physics Letters, 25, 311 (1974).

5. Hudgin , R. and Lipson, S.G., "Analysis of a Monolithic Piezoelectric Mirror", J. Applied Physics 46, (Feb., 1975).

OPTICAL WAVEFRONT CORRECTION IN REAL TIME*

Virendra N. Mahajan
The Charles Stark Draper Laboratory, Inc.
Cambridge, Massachusetts 02139

Abstract

A theory of optical Bragg diffraction is given and it is shown that an array of Bragg cells, each cell carrying two orthogonal sound waves, can be used to correct in real time section-by-section amplitude and phase and overall tilt errors of an optical wavefront distorted by atmospheric turbulence. Because of the tilt correction an object is automatically tracked and the image stabilized. A section-by-section tilt correction can be achieved if two cell arrays are used. The spectral bandwidth of an image corrected by Bragg cells is also considered.

Introduction

When an airborne or a space object is imaged by a ground-based telescope, the atmospheric turbulence, with its associated refractive-index inhomogeneities, distorts the optical wavefronts and thus degrades the image quality. Over a small section of a wavefront the distortion produced at any instant may be approximated by a uniform change in amplitude, a constant phase shift, and a simple tilt. The amount of distortion varies from one section to another, and changes every few milliseconds or faster. Therefore, with a small-aperture telescope a wandering image of changing irradiance is obtained. However, with a large-aperture telescope, which may be considered as being made up of many small ones, a blurred image results. Theoretically, the resolution of the image should increase linearly with the size of the aperture, but because of atmospheric turbulence, it remains constant beyond an approximately 10-cm aperture.

By applying real-time wavefront correction, the image resolution can be improved considerably. We have studied Bragg diffraction of light by sound waves for this purpose.(1,2) Compared to deformable mirrors, which correct for phase errors only,(3) this approach is faster and can control the amplitude, the direction of propagation, and the phase of a wavefront. Furthermore, the absence of moving parts, such as in a segmented mirror, provides a more reliable system. As in any other method, the object must lie in an isoplantic angle of the atmosphere.

In this paper a theory of optical Bragg diffraction is given and a step-by-step approach to correct distorted optical wavefronts in real time is described.

Optical Bragg Diffraction

Raman-Nath Equation

A sound wave causes light diffraction by producing periodic variations of the refractive index of the medium in which it is traveling. As indicated in Fig. 1, consider a plane wave of light of amplitude U, frequency ω, wave vector $\vec{k}$, and phase constant ϕ traveling in a transparent medium of refractive index n. We assume that $\vec{k}$ lies in the zx plane. When a sound wave of frequency Ω, wave vector $\vec{K}$, and phase constant Φ traveling along the x axis with a velocity V is generated in the medium, its refractive index in the region of the sound wave can be written

$$n(\vec{r}, t) = n + \Delta n \sin(\Omega t - \vec{K} \cdot \vec{r} + \Phi) , \quad 0 \le z \le L , \tag{1}$$

where Δn is the amplitude of the refractive-index wave of width L. The vector $\vec{K}$ is either parallel or antiparallel to the x axis. Sound waves are generally generated by attaching a piezoelectric transducer to the medium.

The optical field incident on the sound wave can be written

$$E(\vec{r}, t) = \frac{1}{2} U \exp[i(\omega t - \vec{k} \cdot \vec{r} + \phi)] + c.c. , \quad z \le 0 . \tag{2}$$

The field inside the sound region satisfies the wave equation which, for a nonconducting and nonmagnetic medium, can be written(1)

*Work supported by the U.S. Army under Contract DAAD07-75-C-0034.

$$\nabla^2 E = \frac{1}{v^2}\frac{\partial^2}{\partial t^2} E[1 + 2\frac{\Delta n}{n}\sin(\Omega t - \vec{K}\cdot\vec{r} + \Phi)] , \quad z \geq 0 , \tag{3}$$

where v is the speed of light in the medium. Because of the periodic nature of $n(\vec{r}, t)$, the optical field can be expanded in a Fourier series

$$E(\vec{r}, t) = \frac{1}{2}\sum_{\ell=-\infty}^{\infty} U_\ell(z) \exp[i(\omega_\ell t - \vec{k}_\ell\cdot\vec{r} + \phi)] + c.c. , \quad z \geq 0 , \tag{4}$$

where U_ℓ is the amplitude of the ℓth-order diffracted wave of frequency

$$\omega_\ell = \omega + \ell\Omega \tag{5}$$

and wave vector

$$\vec{k}_\ell = \vec{k} + \ell\vec{K} . \tag{6}$$

The amplitudes satisfy the initial conditions

$$U_\ell(0) = U, \text{ for } \ell = 0 , \quad = 0, \text{ otherwise.} \tag{7}$$

The time-averaged irradiance of a diffracted wave emerging from the sound region is given by

$$I_\ell = \frac{1}{2}|U_\ell(L)|^2 . \tag{8}$$

Substituting Eq. (4) into Eq. (3), neglecting the second derivative of $U_\ell(z)$ with respect to z, and using the fact that $\Omega \ll \omega$, we obtain the following Raman-Nath equation for the diffracted amplitudes

$$\frac{dU_\ell}{dz} + \frac{k^2}{2n}\frac{\Delta n}{k_{\ell z}}[U_{\ell-1}\exp(i\Phi) - U_{\ell+1}\exp(-i\Phi)] = \frac{i}{2k_{\ell z}}(k_\ell^2 - k^2)U_\ell , \tag{9}$$

where $k_{\ell z}$ is the z component and k_ℓ is the magnitude of $\vec{k}_\ell$.

Light Incident From Above the Acoustic Wavefronts

When light is incident from above the acoustic wavefronts as in Fig. 2, the wave vectors of the incident light and sound wave are given by

$$\vec{k} = (-k\sin\theta, 0, k\cos\theta) \tag{10}$$

and

$$\vec{K} = (K, 0, 0) , \tag{11}$$

respectively. Equation (6) can therefore be written

$$\vec{k}_\ell = (-k\sin\theta + \ell K, 0, k\cos\theta) . \tag{12}$$

Substituting Eqs. (10) and (12) into Eq. (9), we obtain

$$\frac{dU_\ell}{dz} + \frac{k\,\Delta n}{2n\cos\theta}[U_{\ell-1}\exp(i\Phi) - U_{\ell+1}\exp(-i\Phi)] = \frac{i\ell K^2}{2k\cos\theta}(\ell - \frac{2k}{K}\sin\theta)U_\ell . \tag{13}$$

In the Bragg region of diffraction,(4) i.e., when $Q = K^2L/k$ is larger than 10, light is diffracted from the zero order, which initially contains all the incident light, into the first order only, provided θ is approximately equal to the Bragg angle Θ where $\sin\Theta = K/2k$. Under these conditions, the set of Eqs. (13) reduces to

$$\frac{dU_0}{dz} - \xi U_1\exp(-i\Phi) = 0 \tag{14}$$

and

$$\frac{dU_1}{dz} + \xi U_0 \exp(i\Phi) = 2i\psi\, U_1 \tag{15}$$

for $\ell = 0$ and $\ell = 1$, respectively, where for small angles

$$\xi = \frac{\Delta n}{n} \frac{k}{2\cos\theta}$$

$$\simeq \pi \frac{\Delta n}{n\lambda} \tag{16}$$

and

$$\psi = \frac{K^2}{4k\cos\theta}\left(1 - \frac{2k}{K}\sin\theta\right)$$

$$\simeq \pi(\Theta - \theta)/\Lambda\ . \tag{17}$$

In Eqs. (16) and (17), λ is the optical wavelength inside the medium and Λ is the acoustic wavelength, respectively. Solving Eqs. (14) and (15) subject to the initial conditions expressed by Eq. (7), the field of the Bragg-diffracted wave, after emerging from the sound wave is found to be

$$E_1(\vec{r}, t) = \frac{1}{2} U_1 \exp[i(\omega_1 t - \vec{k}_1 \cdot \vec{r} + \phi)] + c.c.\ , \tag{18}$$

where its amplitude, frequency, and wave vector are given by

$$U_1 = U(\xi/\zeta)\sin(\zeta L)\exp[i(\pi + \Phi + \psi L)]\ , \tag{19}$$

$$\omega_1 = \omega + \Omega\ , \tag{20}$$

and

$$\vec{k}_1 = (K - k\sin\theta,\ 0,\ k\cos\theta)\ , \tag{21}$$

respectively. The quantity ζ in Eq. (19) is given by

$$\zeta = (\xi^2 + \psi^2)^{1/2}\ . \tag{22}$$

The time-averaged irradiance of the wave is given by

$$I_1 = I(\xi/\zeta)^2 \sin^2(\zeta L)\ , \tag{23}$$

where $I = \frac{1}{2}U^2$ is the time-averaged irradiance of the incident light. The irradiance of the undiffracted light is given by $I_0 = I - I_1$.

Note that whereas the frequency of the undiffracted (zero-order) wave is the same as that of the incident light, the frequency of the diffracted wave is larger by an amount equal to the acoustic frequency. If, however, the traveling sound wave is reflected from the other end of the medium so that a standing sound wave is formed, then both the undiffracted and the diffracted waves are made up of an infinite number of waves whose frequencies are different from the frequency of the incident light by even multiples of the acoustic frequency in the case of undiffracted wave and odd multiples in the case of diffracted wave.[5] For this reason the sound wave should be either absorbed at the other end or reflected at a sufficient angle so that the reflected sound wave does not diffract any light (because it does not satisfy the Bragg condition).

Incident and Diffracted Fields in Free Space

Let us now express the incident and Bragg-diffracted fields in free space using subscripts i and d to identify them. Thus, for example, we denote the amplitude, wave vector, wavelength, and the angle of incidence of the incident light by U_i, $\vec{k}_i$, λ_i, and θ_i, respectively.

When light represented by field

$$\mathcal{E}_i = \frac{1}{2} U_i \exp[i(\omega t - \vec{k}_i \cdot \vec{r} + \phi)] + c.c. , \tag{24}$$

where

$$\vec{k}_i = (-k_i \sin\theta_i, 0, k_i \cos\theta_i) \tag{25}$$

is incident on the Bragg cell, the field of the Bragg-diffracted wave, neglecting absorption and reflection of light by the cell, is given by

$$\mathcal{E}_d = \frac{1}{2} U_d \exp[i(\omega_d t - \vec{k}_d \cdot \vec{r} + \phi + \Phi)] + c.c. , \tag{26}$$

where

$$U_d = U_i(\pi \Delta n/\lambda_i \zeta) \sin(\zeta L) \exp\{i[\pi + (\theta_i - \frac{\lambda_i}{2\Lambda}) \frac{L}{n\Lambda}]\} , \tag{27}$$

$$\vec{k}_d = (k_d \sin\theta_d, 0, k_d \cos\theta_d) , \tag{28}$$

$$\omega_d = \omega + \Omega , \tag{29}$$

$$\zeta = \pi[(\Delta n/\lambda_i)^2 + (\theta_i - \lambda_i/2\Lambda)^2 (1/n\Lambda)^2]^{1/2} , \tag{30}$$

and

$$\theta_d = \frac{\lambda_i}{\Lambda} - \theta_i . \tag{31}$$

The time-averaged irradiance of the diffracted wave is given by

$$I_d = I_i(\pi \Delta n/\lambda_i \zeta)^2 \sin^2(\zeta L) , \tag{32}$$

where $I_i = \frac{1}{2} U_i^2$ is the time-averaged irradiance of the incident light.

From these equations we note that all of the incident light is diffracted if it is incident at the Bragg angle (i.e., if $\theta_i = \lambda_i/2\Lambda$) and if $\Delta n = \lambda_i/2L$. Moreover, the amplitude, direction of propagation, and phase of a Bragg-diffracted wave can be controlled by varying the amplitude, frequency, and phase of the electrical signal generating the sound wave. These properties of the diffracted wave can be used for real-time correction of an optical wavefront distorted, for example, by atmospheric turbulence.

Subwavefront Phase and Overall Tilt Correction

Consider an optical wavefront distorted by atmospheric turbulence. Let us divide this distorted wavefront into small sections (determined by the turbulence cell size) such that each section, referred to as a subwavefront, is planar, and differs from its corresponding undistorted subwavefront by a constant phase shift (mean piston error) and a simple tilt. Owing to the random nature of the atmospheric turbulence, the relative phase and tilt of a subwavefront change independently of the changes in other subwavefronts. If the distorted wavefront is incident on an array of Bragg cells such that each cell intercepts only one subwavefront, the amplitude and phase errors of the subwavefronts can be corrected in real time by changing the amplitudes and phases of the electrical signals driving the Bragg cells. The wavefront errors are determined by performing a modified Hartmann test as described in Reference 1.

It is clear that the tilt error of a subwavefront can be corrected by changing the frequency of the electrical signal. However, if the tilt errors vary significantly from one subwavefront to another, the frequencies of the sound waves required to correct them will differ from cell to cell. Accordingly, the corresponding diffracted subwaves will have different frequencies. Because a frequency shift represents a time-dependent phase, the diffracted subwaves will not form a coherent wavefront. Therefore, either all the sound waves must have the same frequency, in which case an overall (rather than section-by-section) tilt of a wavefront can be corrected, or, we must somehow eliminate the frequency shifts.

Thus if the Bragg cells are driven by electrical signals of the same frequency but controllable relative amplitudes and phases, the subwavefront amplitude and phase errors of an atmospherically distorted wavefront can be corrected. By varying the frequency of the sound waves propagating along the x axis, an overall tilt error of the wavefront in the zx plane can be corrected. An overall tilt error in the yz plane can be corrected if in each cell another sound wave propagating along the y axis is generated.(6) Each cell of the array then consists of two orthogonal sound waves propagating along the x and y axes, and all the x-waves have the same frequency and all the y-waves have the same frequency; the light is incident close to the z axis at their Bragg angles. The xy-diffracted waves, called 2-d waves, have the same frequency, are coherent with each other and form a wavefront corrected for section-by-section phase and overall tilt errors. The frequency of the 2-d waves is different from that of the incident light by the acoustic frequencies, but this difference is negligibly small. Because of the overall tilt correction the object is automatically tracked and the image stabilized.

Elimination of the Frequency Shifts

The frequency shift produced by a sound wave, can be eliminated by rediffracting the diffracted wave by another sound wave of the same frequency in such a way that the frequency shifts produced by the two cancel each other, while maintaining the direction-of-propagation dependence of the diffracted-diffracted (d-d) wave on their frequency. It is clear that we need to find light-sound interaction geometries which produce frequency shifts of sign opposite to the one produced by the geometry of Fig. 2. Two such geometries are possible, one in which light is incident from below the acoustic wavefronts and the other in which the direction of sound propagation is reversed. We first consider light diffraction under these conditions and then by two approximately parallel and antiparallel sound waves.

Diffraction of Light Incident from Below the Acoustic Wavefronts

From Fig. 3 we note that when light is incident from below the acoustic wavefronts, the wave vector of the incident light and sound wave are given by

$$\vec{k} = (k \sin\theta,\ 0,\ k \cos\theta) \tag{33}$$

and

$$\vec{K} = (K,\ 0,\ 0)\ , \tag{34}$$

respectively. The wave vectors of the diffracted waves, Eq. (6), can therefore be written

$$\vec{k}_\ell = (\ell K + k \sin\theta,\ 0,\ k \cos\theta)\ . \tag{35}$$

Substituting Eqs. (33) and (35) into Eq. (9), the Raman-Nath equation for the interaction geometry of Fig. 3 becomes

$$\frac{dU_\ell}{dz} + \frac{k\ \Delta n}{2n \cos\theta}\,[U_{\ell-1} \exp(i\Phi) - U_{\ell+1} \exp(-i\Phi)] = \frac{i\ell K^2}{2k \cos\theta}\,(\ell + \frac{2k}{K} \sin\theta) U_\ell\ . \tag{36}$$

In the Bragg region of diffraction, an examination of Eq. (36) shows that light is now diffracted into the $\ell = -1$ order only, assuming, of course, that $\theta \simeq \Theta$. Thus the set of Eqs. (36) reduces to

$$\frac{dU_0}{dz} + \xi U_{-1} \exp(i\Phi) = 0 \tag{37}$$

and

$$\frac{dU_{-1}}{dz} - \xi U_0 \exp(-i\Phi) = 2i\psi\ U_{-1}\ . \tag{38}$$

Comparing Eqs. (37) and (38) with Eqs. (14) and (15), the field of the Bragg-diffracted wave emerging from the sound region can be written

$$E_{-1}(\vec{r}, t) = \frac{1}{2} U_{-1} \exp[i(\omega_{-1} t - \vec{k}_{-1} \cdot \vec{r} + \phi)] + c.c.\ , \tag{39}$$

where its amplitude, frequency, and wave vector are given by

$$U_{-1} = U(\xi/\zeta) \sin(\zeta L) \exp[i(\psi L - \Phi)]\ , \tag{40}$$

$$\omega_{-1} = \omega - \Omega\ , \tag{41}$$

and

$$\vec{k}_{-1} = (k \sin\theta - K,\ 0,\ k \cos\theta)\ , \tag{42}$$

respectively. The frequency of the diffracted wave is smaller than that of the incident light by an amount equal to the acoustic frequency.

Light Diffraction by a Sound Wave Traveling in the Negative x Direction

From Fig. 4 we note that

$$\vec{k} = (-k \sin\theta,\ 0,\ k \cos\theta) \tag{43}$$

and

$$\vec{K} = (-K,\ 0,\ 0) \tag{44}$$

so that

$$\vec{k}_{\ell} = (-k \sin\theta - \ell K,\ 0,\ k \cos\theta)\ . \tag{45}$$

Substituting Eqs. (43) and (45) into Eq. (9) we obtain Eq. (36). Hence the diffraction of light incident from above the acoustic wavefronts propagating in the negative x direction is identical to that of light incident from below the acoustic wavefronts propagating in the positive x direction.

Successive Bragg Diffraction of Light by Two Approximately Parallel Sound Waves

As indicated in Fig. 5, consider two sound waves of frequency Ω, wave numbers K and K', phase constants Φ and Φ', widths L and L', propagating along the x and x' axes with velocities V and V' in media of refractive-indices n and n', respectively. Let the corresponding refractive-index waves be given by

$$\Delta n(x,\ t) = \Delta n \sin(\Omega t - Kx + \Phi) \tag{46}$$

and

$$\Delta n'(x',\ t) = \Delta n' \sin(\Omega t - K'x' + \Phi')\ , \tag{47}$$

respectively. We assume that the angle between the x and x' axes is equal to the difference between the Bragg angles (inside the media) for the two sound waves. This has the effect that light diffracted by one sound wave at its Bragg angle is incident on the other sound wave at its Bragg angle, resulting in optimum diffraction efficiency.

When light represented by Eq. (24) is incident on the composite Bragg cell, part of it is diffracted by the first sound wave and part of this diffracted wave is in turn diffracted by the second sound wave. The field of the diffracted-diffracted Bragg wave, neglecting absorption and reflection by the cell, is given by (using the results obtained above)

$$(\mathcal{E}_{dd})_{\uparrow\uparrow} = \frac{1}{2}\,(U_{dd})_{\uparrow\uparrow}\, \exp\{i[\omega t - (\vec{k}_{dd})_{\uparrow\uparrow} \cdot \vec{r} + \phi + \Phi - \Phi']\} + c.c.\,, \tag{48}$$

where its amplitude $(U_{dd})_{\uparrow\uparrow}$ and the wave vector $(k_{dd})_{\uparrow\uparrow}$ are given by

$$(U_{dd})_{\uparrow\uparrow} = U_i[(\pi\ \Delta n/\lambda_i \zeta)\ \sin(\zeta L)][(\pi\ \Delta n'/\lambda_d \zeta')\ \sin(\zeta' L')] \cdot \exp\{i\pi[1 - (\theta_i - \frac{\lambda_i}{2\Lambda})(\frac{L}{n\Lambda} - \frac{L'}{n'\Lambda'})]\} \tag{49}$$

and

$$(k_{dd})_{\uparrow\uparrow} = [-k_i \sin(\theta_{dd})_{\uparrow\uparrow},\ 0,\ k_i \cos(\theta_{dd})_{\uparrow\uparrow}]\ , \tag{50}$$

respectively. In Eqs. (49) and (50)

$$\lambda_d = 2\pi c/(\omega + \Omega)\ , \tag{51}$$

$$\zeta' = \pi[(\Delta n'/\lambda_d)^2 + (\theta_i - \lambda_i/2\Lambda)^2(1/n'\Lambda')^2]^{1/2}\ , \tag{52}$$

and

$$(\theta_{dd})_{\uparrow\uparrow} \simeq \theta_i - \frac{\lambda_i}{\Lambda} + \frac{\lambda_d}{\Lambda'} . \qquad (53)$$

The quantity λ_d, given by Eq. (51), is the wavelength of the Bragg-diffracted wave produced by the first sound wave. Because $\Omega \ll \omega$, we may write $\lambda_d = \lambda_i$.

It is evident from these equations that the amplitude, phase, and direction of propagation of the d-d wave can be controlled by varying the amplitudes, relative phase, and the frequency of the sound waves. The d-d wave has the same frequency as the incident light and the two are therefore coherent with each other

Successive Bragg Diffraction of Light by Two Approximately Antiparallel Sound Waves

Consider two sound waves as above except that now they are traveling in approximately opposite directions (Fig. 6), the angle between their directions of propagation being equal to the sum of their Bragg angles (inside the media). The refractive-index waves produced by them can be written

$$\Delta n(x, t) = \Delta n \sin(\Omega t - Kx + \Phi) \qquad (54)$$

and

$$\Delta n'(x', t) = \Delta n' \sin(\Omega t + Kx' + \Phi') . \qquad (55)$$

The field of the d-d wave in this case is given by

$$(\mathcal{E}_{dd})_{\uparrow\downarrow} = \frac{1}{2} (U_{dd})_{\uparrow\downarrow} \exp\{i[\omega t - (\vec{k}_{dd})_{\uparrow\downarrow} \cdot \vec{r} + \phi + \Phi - \Phi']\} + c.c., \qquad (56)$$

where

$$(U_{dd})_{\uparrow\downarrow} = (U_{dd})_{\uparrow\uparrow} , \qquad (57)$$

$$(\vec{k}_{dd})_{\uparrow\downarrow} = [k_i \sin(\theta_{dd})_{\uparrow\downarrow}, 0, k \cos(\theta_{dd})_{\uparrow\downarrow}] , \qquad (58)$$

and

$$(\theta_{dd})_{\uparrow\downarrow} \simeq -\theta_i + \frac{\lambda_i}{\Lambda} + \frac{\lambda_d}{\Lambda'} . \qquad (59)$$

The d-d wave again has the same frequency as the incident light.

Comparison of Successive Bragg Diffraction by Two Approximately Parallel and Antiparallel Sound Waves

The major difference between the properties of the d-d waves produced by diffraction of light by two approximately parallel and antiparallel sound waves lies in their direction of propagation. In the parallel case the d-d wave appears at an angle

$$(\theta_{dd})_{\uparrow\uparrow} = \theta_i + \lambda_i f \frac{V - V'}{VV'} , \qquad (60)$$

where $f = \Omega/2\pi$ (acoustic frequency in Hz) and we have written $\lambda_d = \lambda_i$. When the sound waves are approximately antiparallel, it appears at an angle

$$(\theta_{dd})_{\uparrow\downarrow} = -\theta_i + \lambda_i f \frac{V + V'}{VV'} . \qquad (61)$$

From Eqs. (60) and (61) it is clear that if θ_i changes, θ_{dd} can be held fixed by changing f, the change in f required for a given change in θ_i being larger in the parallel case by a factor $(V + V')/(V - V')$ than in the antiparallel one. For the same reason the antiparallel case requires more stable electrical signals to drive the cells. It should be noted that in the parallel case the two sound waves must propagate with different velocities if $(\theta_{dd})_{\uparrow\uparrow}$ is to be different from θ_i; this restriction does not apply to the antiparallel case where V and V' can be equal.

Another difference between the two cases may be noted from Eqs. (49) and (57), according to which the phase factor

$$(\theta_i - \lambda_i/2\Lambda)(\frac{L}{n\Lambda} - \frac{L'}{n'\Lambda'})$$

in the antiparallel case can be zero for all angles of incidence because in this case Λ and Λ' can be equal. Choosing $L = L'$, $n = n'$ and $\Lambda = \Lambda'$, the phase factor is always zero. This cannot be true in the parallel case.

Phase and Tilt Correction of Subwavefronts

From Eqs. (48) and (56) we note that when light is diffracted successively by two sound waves, the phase of the d-d wave can be controlled by varying their relative phase. According to Eqs. (53) and (59), its direction of propagation in the zx plane can be controlled by varying their frequency. By generating orthogonal sound waves of a given frequency, one traveling along the y axis in the first medium and another along the y' axis in the second medium, the direction and phase of the 4-d wave produced by diffraction of the incident light by all the four sound waves can be controlled for arbitrary tilts and phase changes of the incident light. Thus if a distorted wavefront is incident on two approximately parallel or antiparallel arrays of Bragg cells, each cell carrying two orthogonal sound waves, the phase and tilt errors of subwavefronts can be corrected by changing the relative phases and frequencies of the electrical signals driving the cells. Whereas in the case of parallel arrays two parallel sound waves of a given frequency must travel with different velocities, in the case of antiparallel arrays two antiparallel sound waves can travel with the same velocity and hence the arrays can be identical.

Chromatic Effects

So far we have assumed that the incident light is monochromatic. When light of different wavelengths is incident at a certain angle on a single or composite Bragg cell, the corresponding diffracted or diffracted-diffracted waves appear at different angles according to Eqs. (31), (53), and (59). Thus the Bragg cells behave as a dispersive medium. Consequently, the image formed by 2-d or 4-d waves will have color misregistration. This misregistration can, however, be removed if these waves are diffracted by another Bragg cell, which we call the achromatizing Bragg cell, carrying two orthogonal sound waves. In the case of 2-d waves, the frequencies of these sound waves are the same as those in the cell array. In the case of 4-d waves they are however different by a factor $V_a(V \mp V')/VV'$ where V_a is the acoustic velocity in the achromatizing cell, and minus and plus signs hold for the parallel and antiparallel cases, respectively. The image formed by the 2-d or 4-d waves produced by the achromatizing cell will be achromatic.

While the achromatic image is error free for light of a given wavelength, it is only partially correct for other wavelengths. Neglecting atmospheric dispersion, a wavefront propagating through atmosphere undergoes optical-path-length errors. Therefore, the phase error of a subwavefront will be different for light of different wavelengths, but its tilt error will be independent of wavelength. The corrections made by a Bragg cell, however, are such that the phase correction is independent of the wavelength of the incident light but tilt correction depends on it. From Eqs. (31), (53), and (59) we find that when the phase and tilt of a subwavefront of a certain wavelength are corrected by changing the acoustic phase and frequency, the fractional residual phase and tilt errors of a subwavefront of another wavelength are equal to its fractional wavelength deviation. Thus, for example, if 10% residual phase and tilt errors are acceptable, the spectral bandwidth is 20% of the central wavelength.

Imaging with Real-Time Wavefront Correction

Figure 7 shows a schematic diagram of imaging with real-time wavefront correction. The Bragg-cell array(s) is placed in the exit pupil of the telescope. A stop placed at the focal plane transmits only the 2-d (4-d) image. A relay lens produces an image of the telescope exit pupil onto a lenticular array of lenses. The light collimated by this lens is divided into two parts by a beam splitter. One part is incident on the lens array which forms poorly resolved images of the object on an array of quadrant photodetectors. A computer determines the wavefront errors from the irradiance and centroids of these images, and the electrical signals driving the cells are changed accordingly. There is a 1:1 correspondence between the turbulence cells, Bragg cells, lenticular lenses, and the photodetectors.

The other part of the collimated beam is demagnified and diffracted by an achromatizing Bragg cell. The 2-d waves produced by this cell are focused and a high-resolution achromatic image, approaching the diffraction limit of the telescope aperture, is formed. Note that because of the tilt corrections the object is automatically tracked and the image stabilized.

References

1. Mahajan, V. N., "Real-Time Wavefront Correction through Bragg Diffraction of Light by Sound Waves," J. Opt. Soc. Am. 65:271, 1975.

2. Mahajan, V. N., "Real-Time Correction of Wavefronts Distorted by Atmospheric Turbulence," Imaging in Astronomy WB5-1, June 1975.

3. Cone, P. F. and Feinleib, J., "High-Speed Deformable Mirrors for Wavefront Correction," (abstr.) J. Opt. Soc. Am. 65:1212, 1975.

4. Klein, W. R. and Cook, B. D., "Unified Approach to Ultrasonic Light Diffraction," IEEE Trans. Sonics Ultrasonics SU-14:123, 1967.

5. Mahajan, V. N. and Gaskill, J. D., "Bragg Diffraction of Light by a Standing Sound Wave," Opt. Acta 21:893, 1974; Mahajan, V. N., "Diffraction of Light by Sound Waves of Arbitrary Standing Wave Ratio," J. Appl. Phys. 46:3707, 1975; Mahajan, V. N., "Theory of Acoustooptic Interaction: Standing and Traveling Sound Waves," Wave Electronics, May 1976 (to be published).

6. Mahajan, V. N. and Gaskill, J. D., "Diffraction of Light by Two Orthogonal Sound Waves," J. Appl. Phys. 45:1518, 1974.

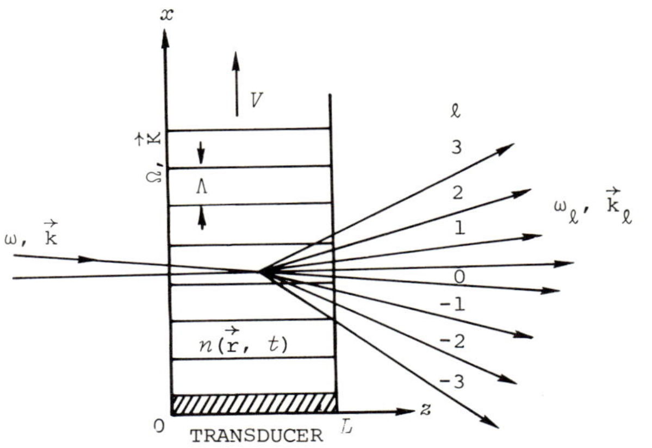

Figure 1. Diffraction of light by a sound wave.

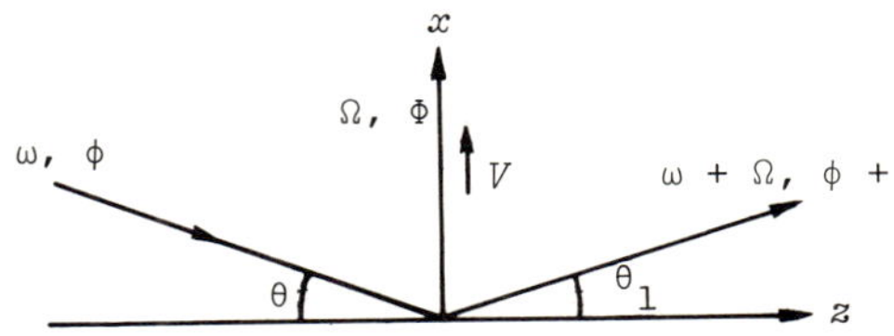

Figure 2. Bragg diffraction of light by a sound wave traveling in the negative x direction. $\theta + \theta_1 \simeq \lambda/\Lambda$.

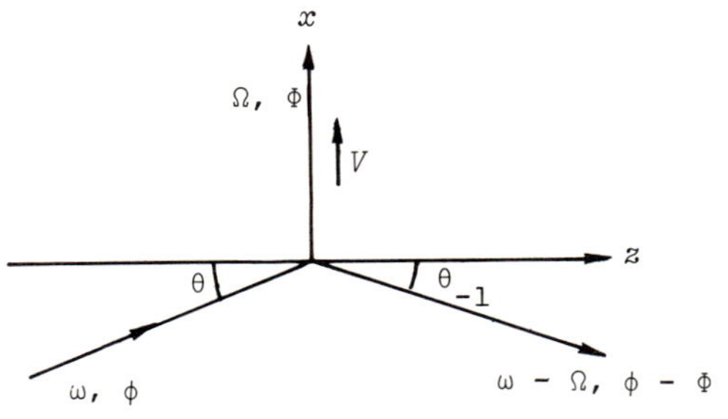

Figure 3. Bragg diffraction of light when incident from below the acoustic wavefronts. $\theta + \theta_{-1} \simeq \lambda/\Lambda$.

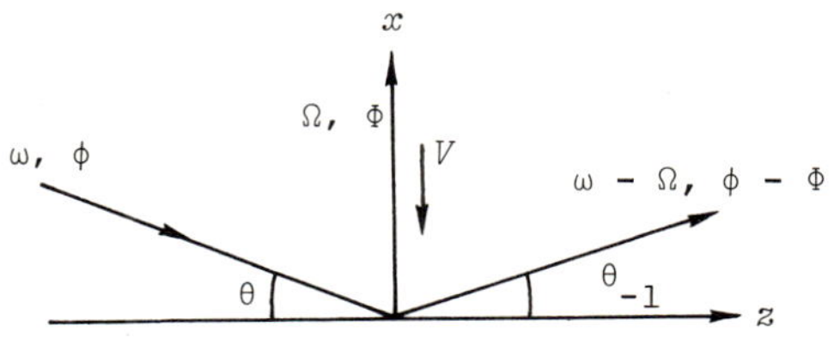

Figure 4. Bragg diffraction of light when incident from above the acoustic wavefronts. $\theta + \theta_{-1} \simeq \lambda/\Lambda$.

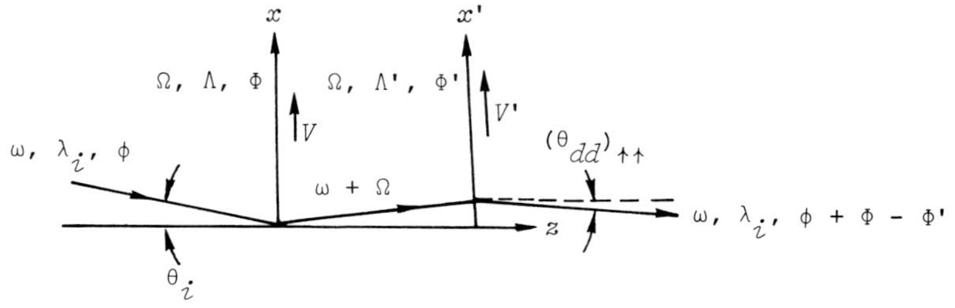

Figure 5. Successive Bragg diffraction of light by two sound waves of the same frequency but different wavelengths. $(\theta_{dd})_{\uparrow\uparrow} = \theta_i + \lambda_i f(V - V')/VV'$.

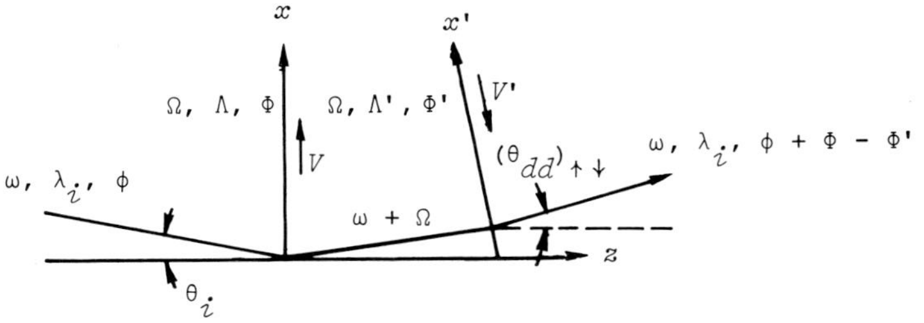

Figure 6. Successive Bragg diffraction of light by two sound waves of the same frequency but not necessarily different wavelengths. $(\theta_{dd})_{\uparrow\downarrow} = -\theta_i + \lambda_i f(V + V')/VV'$.

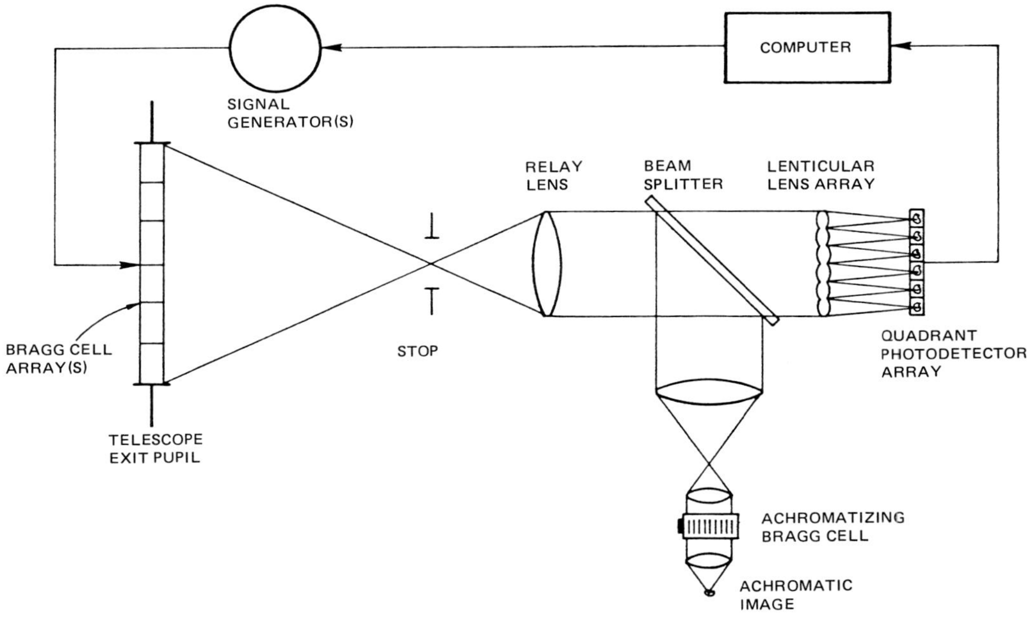

Figure 7. Imaging with real-time wavefront correction.

THE EFFECTS OF ATMOSPHERIC DISPERSION
ON COMPENSATED IMAGING*

Edward P. Wallner
Itek Corporation
Lexington, Mass. 02173

Abstract

Rays from a given point on a celestial object will traverse different paths through the atmosphere due to atmospheric dispersion. They will therefore suffer different phase distortions due to atmospheric turbulence. When the mean distortion is corrected by a compensated imaging system a residual error, which is a function of spectral bandwidth, will remain. This error grows as $\sec^{4/3}\zeta \tan^{5/6}\zeta$, where ζ is the zenith angle, and will limit the region of the sky over which satisfactory compensation is possible.

Introduction

The effects of atmospheric dispersion or variation of the index of refraction of the atmosphere with wavelength are usually negligible in astronomical work except at large zenith angles. In that case the effects may be corrected by use of a prism with compensating dispersion.

The problem is more complex for compensated imaging when the phase distortions produced by atmospheric turbulence in the wavefronts from an object are to be removed by a wavefront corrector. Since rays of different color from the same point on the object traverse different atmospheric paths, they will undergo different phase distortions, and cannot all be corrected by the same wavefront corrector.

The resulting error is analyzed below and evaluated for a specific turbulence model.

Atmospheric Refraction

The notation used is illustrated in Figure 1.

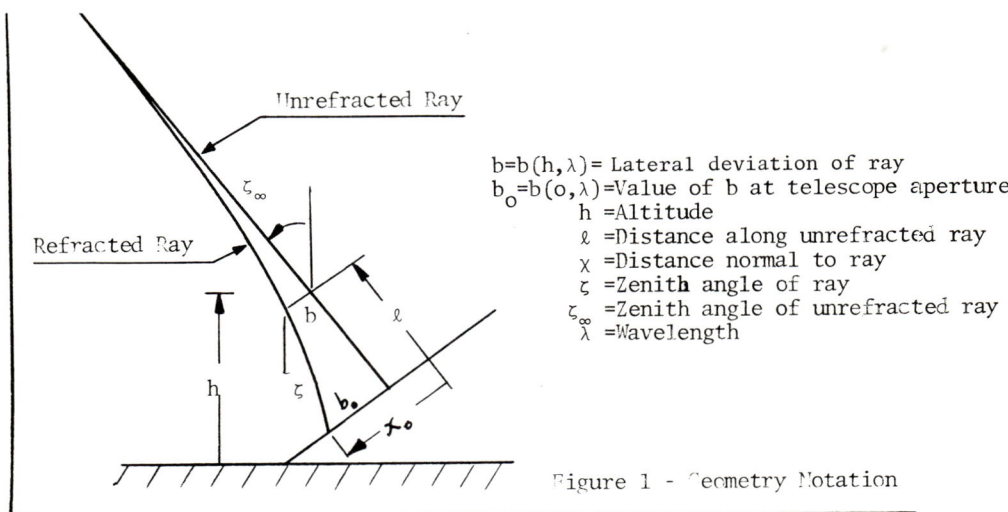

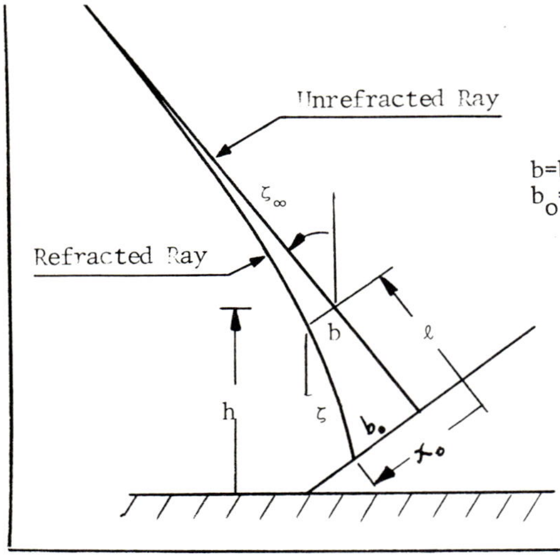

$b=b(h,\lambda)$ = Lateral deviation of ray
$b_o=b(o,\lambda)$ = Value of b at telescope aperture
h = Altitude
ℓ = Distance along unrefracted ray
χ = Distance normal to ray
ζ = Zenith angle of ray
ζ_∞ = Zenith angle of unrefracted ray
λ = Wavelength

Figure 1 - Geometry Notation

References

*This paper is based on research done for AFSC, Rome Air Development Center under Contract F30602-76-C-0053.

A ray from a distant object at true zenith angle ζ_∞ is refracted by the atmosphere and deviates from the unrefracted ray in position and angle. Assuming a horizontally stratified atmosphere and a flat earth, Snell's law may be written as:

$$n(h,\lambda)\sin\zeta = n_\infty \sin\zeta_\infty \qquad (1)$$

where $n(h,\lambda)$ = Index of refraction at altitude h, wavelength λ.

Since the ray is assumed to originate beyond the atmosphere, $n_\infty = 1$.

Differentiating (1) and rearranging:

$$d\zeta = -\frac{dn}{n} \tan\zeta \qquad (2)$$

The index of air is little different from unity over the visible band and the maximum angle of deviation is about 1/2 degree, so that (2) may be closely approximated as

$$\zeta-\zeta_\infty = -\frac{n-n_\infty}{n_\infty} \tan\zeta_\infty = -(n-1)\tan\zeta_\infty \qquad (3)$$

The variation of index n or refractivity N with wavelength can be represented by

$$N_s(\lambda) = n_s(\lambda)-1 = 6.4328 \times 10^{-5}$$

$$+ 2.949810 \times 10^{-2}/(146-\lambda^{-2}) + 2.5540 \times 10^{-4}/(41-\lambda^{-2}) \qquad (4)$$

where λ = wavelength in μm and subscript s indicates standard temperature and pressure. (See Reference 1, pg-111)

For an atmosphere with fixed composition (which is a good description of the troposphere) the index of refraction is function only of density, which is in turn a function of altitude.

$$N(h,\lambda) = n(h,\lambda)-1 = N_s(\lambda) \frac{\rho(h)}{\rho_s}. \qquad (5)$$

where $\rho(h)$ = Atmospheric density at altitude h
ρ_s = Standard density = 1.2250×10^{-3}g cm^{-3}

(See Reference 2, p 88)

The values of zenith angle ζ and lateral displacement from the refracted ray b can be found by substituting (5) into (3) and integrating

$$\zeta-\zeta_\infty = -N_s(\lambda)\tan\zeta_\infty\, \rho(h)/\rho_s$$

$$b(h,\lambda) = \int_\infty^{h\sec\zeta_\infty} dl\, (\zeta-\zeta_\infty) = \sec\zeta_\infty \int_\infty^{h} dh\, (\zeta-\zeta_\infty) \qquad (6)$$

$$= N_s(\lambda)\sec\zeta_\infty \tan\zeta_\infty \frac{1}{\rho_s}\int_h^\infty dh\, \rho(h) \qquad (7)$$

Assuming a flat earth, the integral in (7) is the atmospheric pressure at altitude h scaled by the acceleration of gravity

$$b(h,\lambda) = N_s(\lambda) \sec\zeta_\infty \tan\zeta_\infty \frac{P(h)}{g\rho_s} \qquad (8)$$

The value of b at any altitude can be compactly expressed in terms of the value at the site altitude and the ratio of pressure at altitude to pressure at the site. Using subscript 0 for site values,

$$b(h,\lambda) = b_0(\lambda)\, P(h)/P_0 \qquad (9)$$

where $$b_0(\lambda) = N_s(\lambda) \sec\zeta_\infty \tan\zeta_\infty (P_0/g\rho_s) \qquad (10)$$

The term $P_0/g\rho_s$ can be interpreted as the effective depth of the atmosphere at the site.

This expression with the values in equations (4) can be used to determine the path of any ray through the atmosphere.

Optical Path Length Distortion

Consider the problem of telescopic viewing of a celestial object from within the atmosphere. Turbulence in the atmosphere causes random local deviations of the refractivity from the average value, which in turn causes random differences in the total optical path length from the object to the telescope aperture. The path length distortion at a given wavelength is proportional to the refractivity at the wavelength, from which only varies by 3% from .4μm to .75μm. In compensated imaging the path length distortion is measured and corrected with an active optical system, e.g., a deformable mirror. If the phase corrector is achromatic, the same path length correction will be made for all colors, which will leave only a small residual error if all rays have followed the same path through the atmosphere, as in the special case of vertical viewing.

In the general case, the optical path length distortion introduced on the ray of wavelength λ arriving at point x in the aperture is the net result of the refractive index disturbance distributed along the entire path, which depends on the lateral position of the ray along the path.

In order to assess the severity of the distortions, it is convenient to express them in terms of wavelength of some arbitrary mid wavelength λ_o.

$$x(h) = x_o - b_o \left[1-\frac{P}{P_o}\right] \qquad (11)$$

$$\phi(x_o,\lambda) = \int_0^\infty d\ell \; \phi_a(h,x) \qquad (12)$$

where $x(h)$ = Lateral position of ray at altitude h

x_o = Lateral position of ray at wavefront corrector

$\phi(x_o,\lambda)$ = Phase distortion at wavefront corrector in waves @ λ_o

$\phi_a(h,x)$ = Index Disturbance at height h, lateral position x in waves m^{-1} @ λ_o

ℓ = Path length along ray = h $\cos\zeta_\infty$

The average phase distortion at point x_o in the corrector is the intensity weighted mean over wavelength of $\phi\ (x_o,\lambda)$

$$\bar{\phi}\ (x_o) = \int_0^\infty d\lambda \; I(\lambda)\ \phi(x_o,\lambda) \qquad (13)$$

where $\bar{\phi}(x_o)$ = Average phase distortion at x

$I(\lambda)$ = Normalized intensity at wavelength λ i.e., $\int_0^\infty d\lambda\ I(\lambda)=1$

The wavefront corrector will correct the mean phase distortion, leaving a phase error of

$$\varepsilon_\phi\ (\lambda) = \phi(x_o,\lambda) - \bar{\phi}(x_o) \qquad (14)$$

This phase error is a random variable dependent on the random phase disturbance. Since the mean value of the phase disturbance is zero, the mean value of the residual error will also be zero. The variance of the error is then the appropriate measure of performance.

In order to evaluate the phase variance, the error in equation (14) will be rewritten using (12) and (13)

$$\varepsilon_\phi(\lambda) = \int_0^\infty d\lambda' \quad I(\lambda') \left(\phi(x_c,\lambda) - \phi(x_c,\lambda')\right)$$

$$= \sec\zeta_\infty \int_0^\infty d\lambda' \int_0^\infty dh \;\; I(\lambda') \left(\phi_a(h,x)-\phi_a(h,x')\right) \qquad (15)$$

where

$$x = x_o - b_o(\lambda) + b_o(\lambda)\ P(h)/P_o$$
$$x' = x_o - b_o(\lambda') + b_o(\lambda')\ P(h)/P_o$$

The square of this error weighted by intensity and averaged over both the spectrum and the ensemble of wavefronts is the desired performance measure.

$$\left\langle \varepsilon_\phi^2 \right\rangle = \int_0^\infty d\lambda\ \ I(\lambda) \left\langle \varepsilon_\phi^2\ (\lambda) \right\rangle$$

$$= \sec^2\zeta_\infty \int_0^\infty d\lambda\ d\lambda'\ d\lambda''\ I(\lambda)\ I(\lambda')\ I(\lambda'') \qquad (16)$$

$$\int_0^\infty dh\ dh' \left\langle \left(\phi_a(h,x) - \phi_a(h,x')\right) \left(\phi_a(h',x) - \phi_a(h',x'')\right) \right\rangle$$

The covariance of the index disturbance at two points in the atmosphere is non-zero only over a region small compared to the total path length. The term in angular brackets in (14) can therefore be expressed as:

$$C(x,\ h,\ x',h') = \left\langle \left(\phi_a(h,x) - \phi_a(h,x')\right)\left(\phi_a(h',x) - \phi_a(h',x'')\right)\right\rangle$$
$$= \left\langle \left(\phi_a(h,x) - \phi_a(h,x')\right)\left(\phi_a(h,x) - \phi_a(h,x'')\right)\right\rangle \delta(\ell-\ell') \qquad (17)$$

where $\delta(\ell)$ is the Dirac delta function and ℓ is path length along the ray.

The delta function may be expressed in terms of altitude instead of path length via the relationship

$$\delta(\ell-\ell') = \delta(h\sec\zeta - h'\sec\zeta) = \cos\zeta\ \delta(h-h') \qquad (18)$$

The expectation term in (16) may be expressed in terms of the index disturbance structure function, which is defined by

$$D_a(h,x-x') = \left\langle \left(\phi_a(h,x) - \phi_a(h,x')\right)^2\right\rangle \qquad (19)$$

Rearranging 17 and substituting 19 gives

$$C(x,\ h,\ x,'h') = \left\langle \frac{1}{2}\left\{\left(\phi_a(h,x) - \phi_a(h,x'')\right)^2 + \left(\phi_a(h,x') - \phi_a(h,x)\right)^2 - \left(\phi_a(h,x') - \phi_a(h,x'')\right)^2\right\}\right\rangle \delta(h-h')\cos\zeta \qquad (20)$$
$$= \frac{1}{2}\left\{D_a(h,x-x'') + D_a(h,x-x') - D_a(h,x'-x'')\right\} \delta(h-h')\cos\zeta$$

For an atmosphere with a Kolmogorov distribution of turbulence, the index structure function may be expressed

$$D_a(h,x-x') = \frac{2.91}{\lambda_o^2}\ C_n^2(h)\,|x-x'|^{5/3} \qquad (21)$$

where D_a is expressed in waves^2 of phase variance at wavelength λ_o.

Substituting (21) and (20) in (16) and integrating over h' gives

$$\left\langle \epsilon_\phi^2 \right\rangle = \sec\zeta_\infty \int_0^\infty d\lambda\ d\lambda'\ d\lambda''\ I(\lambda)\ I(\lambda')\ I(\lambda'')$$
$$\frac{1}{2}\int_0^\infty dh\ \frac{2.91}{\lambda_o^2}\ C_n^2(h)\left\{|x-x''|^{5/3} + |x-x'|^{5/3} - |x'-x''|^{5/3}\right\} \qquad (22)$$

Substituting for x as defined following Equation (16)

$$\left\langle \epsilon_{\phi}^{2} \right\rangle = \sec \zeta_{\infty} \int_{0}^{\infty} d\lambda\, d\lambda'\, d\lambda''\, I(\lambda)\, I(\lambda')\, I(\lambda'')$$

$$\frac{1}{2} \int_{0}^{\infty} dh\, \frac{2.91}{\lambda_{0}^{2}}\, C_{n}^{2}(h)\, |1-P(h)/P_{0}|^{5/3} \Big\{ |b_{0}(\lambda)-b_{0}(\lambda'')|^{5/3}$$

$$+ |b_{0}(\lambda)-b_{0}(\lambda')|^{5/3} - |b_{0}(\lambda')-b_{0}(\lambda'')|^{5/3} \Big\}$$

$$= \frac{1}{2} \sec \zeta_{\infty} \int_{0}^{\infty} d\lambda\, d\lambda'\, I(\lambda)\, I(\lambda')\, |b_{0}(\lambda)-b_{0}(\lambda')|^{5/3}$$

$$\int_{0}^{h} dh\, \frac{2.91}{\lambda_{0}^{2}}\, C_{n}^{2}(h)\, |1-P(h)/P_{0}|^{5/3} \qquad (23)$$

Finally, substituting the definition of $b_0(x)$ from Equation (8)

$$\left\langle \epsilon_{\phi}^{2} \right\rangle = \frac{1}{2} \sec^{8/3} \zeta_{\infty} \tan^{5/3} \zeta_{\infty} \left(P_{0}/g\rho_{s}\right)^{5/3} \qquad (24)$$

$$\int_{0}^{\infty} d\lambda\, d\lambda'\, I(\lambda)\, I(\lambda')\, |N_{s}(\lambda) - N_{s}(\lambda')|^{5/3} \int_{0}^{\infty} dh\, \frac{2.91}{\lambda_{0}^{2}}\, C_{n}^{2}(h)\, |1-P(h)/P_{0}|^{5/3}$$

This is the desired expression for the residual phase variance due to atmospheric dispersion. The variance increases steeply with zenith distance, and the effects of intensity distribution and turbulence distribution appear in separate integrals.

Example

The dispersion effect analyzed here has been evaluated for the case of planetary observation from an observatory at sea level.

The atmospheric model used is the ICAO Standard Atmosphere described in reference 3. This atmosphere has a sea level pressure of 1013 millibars, a sea level temperature of 15° C and a lapse rate of 6.5° C per Km.

The turbulence model used is adapted from model 3 in reference 4 with the low altitude portion of the model being replaced by a purely exponential term.

$$C_{n}^{2}(h) = C_{1} e^{-h/h_{s}} + C_{2} \frac{1}{\sqrt{2\pi}\, w} e^{-\frac{(h-h_{t})^{2}}{2w^{2}}} \qquad (25)$$

where C_1, C_2 = constants
h_s = scale height of low altitude turbulence
h_t = height of tropopause
w = width of tropopause layer

The parameter values used are
h_s = 3200m

h_t = 15000m
w = 700m.

(The value of h_s is suggested by the exponential factor in the model used by Fried and Cloud in Ref. 5. The other values are from Reference 4.)

Two turbulence conditions were evaluated. In the first, there was no high altitude layer at the tropopause and the value of r_0 was 10cm at a wavelength of .55μm. The corresponding value of C_1 is 2.628 x 10^{-16} $m^{-2/3}$.

In the second, r_0 is again 10cm but the tropopause layer contributes half the total turbulence. Here $C_1 = 1.314 \times 10^{-17} m^{-2/3}$ and $C_2 = 4.204 \times 10^{-13} m^{+1/3}$.

The results of numerical integration of the altitude integral in (24) are 1.30 and 3.95 for the two turbulence conditions. The equivalent depth of the atmosphere, $P_o/\rho_s g$ is 8435 meters, giving 3.49 x 10^6 for the multiplier $(P_o/\rho_s g)^{5/3}$ in equation (24).

The integral over wavelength in equation (24) has also been evaluated numerically for the intensity distribution shown in Figure 2.

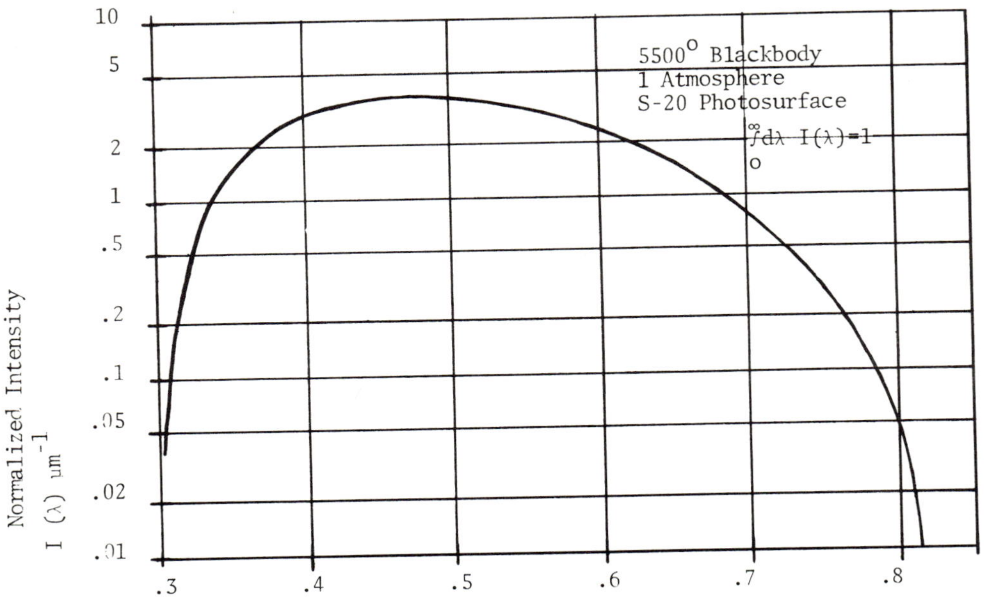

Wavelength μm

Figure 2 Intensity vs Wavelength

The intensity corresponds to radiation from a 5500°K blackbody, corresponding roughly to solar illumination on a grey planetary surface, as transmitted by the earth's atmosphere and detected by an S-20 photosurface. (The atmospheric transmission and photo response are taken from reference 6, Figures 7-1 and 10-5. The value for an air mass of unity was used. Though this strictly applies only to a zenith distance of 0°, the spectral distribution is a reasonable approximation over the region of interest here.)

The resulting value for the integral with this intensity distribution and the index of refraction given by Equation (4) is:

$$G(\lambda) = \int_0^\infty d\lambda\ d\lambda'\ I(\lambda)\ I(\lambda')\left|N_s(\lambda) - N_s(\lambda')\right|^{5/3} = 7.06\times10^{-10} \quad (26)$$

If the radiation is cut off sharply at 0.4 and 0.75 μm, the value of this integral is reduced to 4.19 x 10^{-10}.

The mean square error in phase for the .4 to .75μm spectrum of Figure 2 and the turbulence model with the tropopause layer is

$$\left\langle \varepsilon^2_\phi \right\rangle = \frac{1}{2} \sec^{8/3} \zeta_\infty \tan^{5/3} \zeta_\infty\ (8435)^{5/3}\ 4.19 \times 10^{-10} \times 3.95$$

$$= 2.89 \times 10^{-3} \sec^{8/3} \zeta_\infty \tan^{5/3} \zeta_\infty \quad (27)$$

The corresponding coefficient with no tropopause layer is 9.51 x 10^{-4}.

The rms phase errors are therefore

$$\varepsilon_{\phi\,rms} = 5.38 \times 10^{-2} \sec^{4/3}\zeta_\infty \tan^{5/6}\zeta_\infty \text{ with tropopause layer}$$

$$= 3.08 \times 10^{-2} \sec^{4/3}\zeta_\infty \tan^{5/6}\zeta_\infty \text{ without tropopause layer}$$

These functions are plotted in Figure 3 as a function of zenith angle.

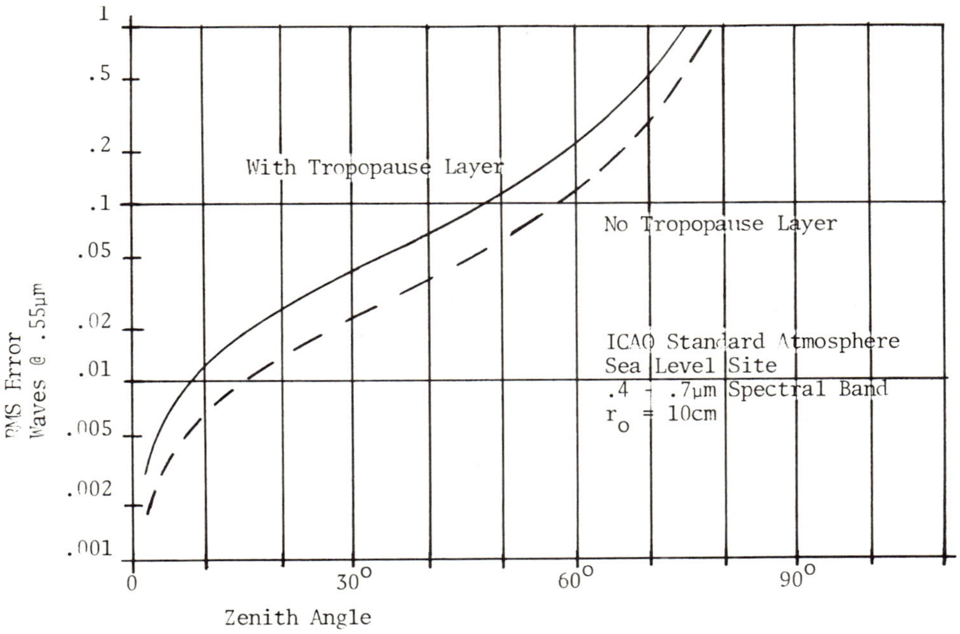

Figure 3 Error vs Zenith Angle

The residual error is seen to grow to excessive values at large zenith angles. This will limit the usable region of the sky to a region about the zenith, the size of which depends on the wavefront quality criterion used.

By restricting the wavelength band employed, this error can be reduced, at the sacrifice of light intensity in the image formed. The steepness of the curves of Figure 3 makes it difficult to enlarge the usable zenith angle greatly.

References

1. American Institute of Physics Handbook, 3rd Ed. McGraw Hill, New York (1972)
2. Born, M and Wolf, E. "Principles of Optics" 3rd Ed. (1965) Pergamon Press, Oxford
3. Manual of the ICAO Atmosphere, NACA TN 3182, May 1954
4. Titterton, P.J. "Scintillation and Transmitter - Aperture Averaging over Vertical Paths" J. Opt. Soc. of Am. v 63 #4 pp 439-444 (1973)
5. Fried, D. L. and Cloud, J. D. "Propagation of an Infinite Plane Wave in a Randomly Inhomogeneous Medium" J. Opt. Soc. of Am. v 56 #12 pp 1667 - 1676 (eq 302)
6. RCA Electro Optics Handbook, First Edition RCA Burlington, Mass. 1968

ADAPTIVE OPTICS FOR SPACE TELESCOPES

T. R. O'Meara, C. J. Swigert and W. P. Brown
Hughes Research Laboratories
Malibu, California 90265

Abstract

Adaptive optics systems for ground-based imaging through the earth's atmosphere must generally measure and correct for path distortions within a time period ranging from 0.5 to 10 milliseconds. For most astronomical objects, this requires a wavefront error sensor of the highest sensitivity — typically a system which employs hundreds of quantum-limited photodetectors devoted exclusively to this task. This paper will point out how the system problems for figure control of orbiting telescopes are quite different since the error sources have periods which typically range from hours to years. Thus error signal integration times can be thousands of times larger for the orbiting optics, and it is feasible and economically advantageous to use low sensitivity dither adaptive optics systems, employing single detectors at the image plane. We will compare three classes of dither systems for this application: 1) one-element-at-a-time step systems; 2) half-at-a-time step systems; and 3) parallel sinusoidal dither (multidither) systems. Several types of signal processing will be compared from a signal-to-noise viewpoint. Computer simulations will be employed to illustrate the system performance at marginal signal-to-noise ratios.

Introduction

We will be discussing the application of adaptive optics to orbiting astronomical telescopes for stellar imagery. We assume that the telescope is sufficiently large and/or lightweight that proper figure control can be maintained only with adaptive optics. Various "on-board" local figure sensing and correction techniques have been proposed[1-5] for this application which, for example, probe a primary mirror figure with a laser test source. However, we feel that systems that measure the actual imaging performance with realistic sources, i.e., stars, have much to recommend them in reduced system costs, achievement of high reliability and overall performance confidence.[8,9]

Fundamentally, we will be considering adaptive optical systems that are imaging unresolved test sources — stars — upon single detectors of diffraction-limited size, or somewhat smaller. We presume these detectors to be elements of a normal imaging tube or discrete detector matrix. The adaptive optical systems to be considered all fall in the category of "hill-climbing" servo mechanisms that maximize the image intensity falling on a single selected detector.

We would next like to point out some key differences between this adaptive optics application and ground-based compensated imaging. The first major difference is in the required servo time constants (integration). Typical time constants for ground-based compensation run on the order of two to ten milliseconds while, for space imaging, time constants run from hours (solar incidence changes) to weeks (mechanical drifts and relaxations). Thus we expect servo integration times of the order of 4×10^5 greater in the space application. This means in turn that one can work with stellar reference or calibration sources that are 4×10^5 weaker than with ground-based imaging. Further, it permits one to work with dithers or trial perturbations on the adaptive mirror actuators that in no way strain the required speed of response, which is not always the case with this class of ground-based systems.

An important secondary consequence is that it permits one to employ the very simple systems of the type to be described. More specifically, this simplicity is not obtained without sacrifice in sensitivity compared to the better interferometric systems, and it is the long integration times that in fact permit operation with moderately insensitive systems.

However, even with the long available integration times, it is still important to operate with a system design that achieves the maximum sensitivity subject to the constraint of employing normal image plane information. This will, in fact, be the major theme of our discussion. As we will see, there are many ways of organizing such systems and some achieve better sensitivity than others.

System Organizations

All of the systems to be considered operate by making a series of trial perturbations on the system actuators.* The information contained in the associated intensity variations, at the test detector, is employed to effect a hill-climbing algorithm that maximizes the intensity.

The actuators may be either employed to effect position and tilt control of well-figured segments and/or may be employed to deform local regions of a deformable mirror. However, it is easiest to understand control algorithms for "piston" mirrors, and these will form the initial basis for our discussion.

The type of perturbation is the logical basis for system characterization. We will be considering both continuous (sinusoidal) perturbations and discrete step perturbation systems. In either case, one can consider perturbations sequenced on single actuators (one-at-a-time) or on blocks of actuators as a whole. In general, it is found that single actuator systems achieve inferior sensitivities (signal-to-noise) simply because the signal (intensity variation) produced by moving a single actuator is small while, at the same time, the background intensity from the unmodulated elements generates a large shot noise component or, in other terms, variance in the measurement process. It can be shown that the signal-to-noise ratio for such systems is inversely proportional to the square of the number of actuators N_a.

Another important way of characterizing these systems is by the way in which the detector output information is used. Specifically, it can be used immediately (or nearly so), which we call "correct-as-you-go," or it can be accumulated with a full set of actuator corrections being estimated as an entity and applied at the end of the dither cycles, which we call "deferred-correction." This choice is essentially independent of the dither process. Examples in all categories will be presented in the following paragraphs.

Sinusoidal Multidither Systems

One possible imaging system, known as a tagging or sinusoidal multidither system, is illustrated in Fig. 1. In the tagging approach, each element is pathlength-modulated with a small phase modulation (typically $\Delta\ell \leq \lambda/6$) by a set of temporally orthogonal waveforms. These are most commonly sinusoids at separate frequencies, one characteristic frequency per element (or in some cases one every two elements). Such systems lend themselves very naturally to analog computer control, as illustrated in Fig. 1. Each dither frequency appears as an amplitude modulation in the output of the detector and is separately extracted by synchronous detection to provide error signals for closed-loop analog servo systems. In the terminology of the previous section, these are "correct-as-you-go" control systems.

These systems are closely similar to the multidither systems that have been successfully applied to the correction of wavefront distortions associated with laser transmitting systems and that have become generally known as Coherent Optical Adaptive Techniques (or COAT**).[6]

The fundamental basis of operation of the multidither correction system to be described here is the fact that the irradiance at the position of a point source image is maximized when the pathlength errors associated with mirror aberrations (in the applications discussed here, the errors introduced by the external paths are negligible) are removed by actuator displacements. In mathematical terms, we wish to maximize the detector irradiance

$$I_o = I(\bar{\phi}_1,\ldots,\bar{\phi}_{N_a}) \tag{1}$$

with respect to the actuator positions $\bar{\phi}_m$ $(m = 1,N_a)$. In the sinusoidal multidither approach the actuators are dithered sinusoidally about their nominal positions, such that

$$\bar{\phi}_a = \phi_a + \psi_o \sin\omega_a t \tag{2}$$

---

*This basic principle, insofar as is known, was first proposed by one of the authors (T. R. O'Meara) at the 1971 Fall Meeting, Scientific Radio Union (URSI).

**We have sometimes referred to the corresponding imaging techniques in a generalized sense as I-COAT for Imaging COAT. In these applications the "coherence" refers, of course, to spatial coherence.

where ψ_o is the dither amplitude and ω_a is the ath dither frequency. The output of a point detector placed at the location of the point source image can be expanded in a Taylor series about the nominal positions ϕ_a, which gives

$$I(\overline{\phi}_m) = I(\phi_m) + \psi_o \sum_{a=1}^{N_a} \frac{\delta I}{\delta \phi_a} \sin\omega_a t + \psi_o^2 \sum_{a=1}^{N_a} \frac{\delta^2 S}{\delta \phi_a^2} \sin^2 \omega t$$

$$+ \psi_o^2 \sum \text{cross-partials} + \text{higher order terms} \quad (3)$$

Note that the coefficients of the linear sinusoidal terms $\sin\omega_a t$ in Eq. (3) are simply the derivatives of the irradiance I_o with respect to the nominal actuator positions. Hence, a servo system that drives the modulation coefficient of the ath actuator dither frequency to zero is also effectively driving the associated partial derivative to zero and satisfying the first order requirement for a maximum in I_o. Moreover, a point with deformable mirrors is that this operation is independent of any knowledge of the mirror deformation shapes or influence functions; the influence function is inherently included.

From a formal viewpoint, we have only satisfied the conditions for a stationary point — not necessarily a maximum. However, in practice, a servo that drives the first order partials to zero, with the correct feedback, produces a maximum. Second order partial derivative information can, however, be useful in some cases.

For specific mirror systems, it is, of course, feasible to derive more explicit results. Thus, for a piston mirror system, the detector output at the ath actuator dither frequency is proportional to

$$I_a = [2J_o(\psi_o)\ (N_a - 1 + H)^{1/2} \sin(\phi_a - \phi_{ac})]\ 2J_1(\psi_o) \sin\omega_a t \quad (4a)$$

where $\phi_a - \phi_{ac}$ is the element phasing error; ϕ_{ac} is the composite phase of the remaining $(N_a - 1)$ elements (excluding a), J_1 is the first order Bessel function and, for small phasing errors,

$$\hat{\phi}_{ac} \doteq \frac{\sum \phi_a}{N_a - 1} \quad (4b)$$

The parameter N_a is the number of actuators and H is a convergence factor

$$H = \sum_{\substack{k,\ell=1 \\ k\neq\ell}}^{N_a} \cos(\phi_k - \phi_\ell) \quad (4c)$$

which ranges between 0 (initial condition with random phasings) and $N_a^2 - N_a$ (when all $\phi_k = \phi_\ell$, i.e., perfect convergence is achieved).

Further, it can be shown that the sign at the second harmonic of the dither frequency is proportional to $\cos(\phi_a - \phi_{ac})$. With both sets of information, one can extract good error estimates for all actuators and thereby achieve a deferred rather than a correct-as-you-go control system.

Servo signal-to-noise ratio is of the form

$$\frac{S}{N} = \left(\frac{\eta\ \phi_e\ T}{N_a}\right)\left(\frac{N_a - 1}{N_a}\right) J_o^2(\psi_o)\ J_1^2(\psi_o) \quad (5a)$$

where T is the servo integration time and ϕ_e is the element phasing error ($\phi_e = \phi_a - \phi_{ac}$). The Bessel function product maximizes for $\psi_o = 1.1$ rad, giving

$$J_o^2(\psi_o)\ J_1^2(\psi_o) = 0.115 \quad (5b)$$

Note that while reasonably large values of ψ_o are desirable for a high S/N, they are detrimental to the imaging process since large dithers disturb the time-averaged telescope figure. One solution is to cycle, alternating between refiguring periods and quiescent imaging periods.

Another "time-sharing" possibility is to employ a synchronized dither technique, reminiscent of laser mode locking. If one applies equally spaced dither frequencies that are phase-locked (synchronized), then, as Fig. 2 illustrates, one obtains repetitive intervals when all elements are momentarily in or nearly in phase; the associated telescope point spread function is excellent. The image detector plane is read out (that is, repetitively exposed) at these intervals only. As a slight variant, one may stop the dither process and hold the actuators at the peak condition for extended times This decreases the required exposure time at the cost of small loss in servo S/N. The cited photoelectron count illustrates the minimum that yielded reasonable performance over a series of computer simulations with this system.

Polystep Dither Systems

In polystep systems, one attempts to bring either one portion of the mirror surface, considered here to be segmented, or an aggregation of elements into phase coincidence, or more accurately pathlength coincidence, with the remainder of the elements. In the one-element-at-a-time approach, each element is "aligned" to the phaser composite from the other elements, starting with any arbitrary element. This process is then sequenced through all elements and recycled, if desired. By "alignment" we mean the process of minimizing the optical path differences (OPD's) from each mirror element to the test detector (with the test star on boresight). It is simplest conceptually to show the OPD's as "phase" errors and to represent the composite field intensities as single wavelength phasor diagrams. The operation of bringing one elemement at a time into phase coincidence with the remainder is illustrated in Fig. 3(a).

A better approach to polystep algorithms is to cycle N/2 elements at a time. A random* set of N/2 elements is brought into phase coincidence with the complementary set of N/2 elements. The second iteration employs a new division into element grouping, and the iterations are repeated until every pairing in an orthogonal set is generated. Hadamard matrices may be employed to generate the element pairings. Figure 3(b) illustrates a convergence sequence with four elements and the same initial conditions as illustrated in Fig. 3(a).

To determine the necessary element phasing, we utilize the fact that piston mirrors produce a cosinusoidal dependence of detector output on element phasing $\phi(x)$. Three trial step perturbations corresponding to a $\phi(x)$ of 0, $\pi/2$, and π radians, yield three outputs I_1, I_2, and I_3. From these outputs, we compute the required element phasing $\phi(x)$ from the equations

$$A \cos\phi = \frac{1}{2} (I_1 - I_3) \tag{6a}$$

$$A \sin\phi = \frac{1}{2} (I_1 - 2I_2 + I_3) \tag{6b}$$

We call such a dither cycle and computation an arc-tan estimator for self-evident reasons.

An alternate procedure is to make dither perturbation steps of $\pm\pi/2$. If the associated intensities are I_2 and I_1, then the associated estimator will be called an arc-sine estimator. For the N/2 dither policy the arc-sine estimator is simply

$$\hat{\phi}_2 = \sin^{-1} \left[\frac{I_2 - I_1}{I_2 + I_1}\right] \tag{7}$$

Although this estimator converges somewhat more slowly with a large initial error than with the arc-tan estimator, it has a much higher S/N ratio. Specifically for the N/2 policy, it is 21/6 times better than the arc-tan estimator. This suggests that one commence system operation with the arc-tan estimator and switch to an arc-sine estimator as convergence is approach — the region where noise becomes troublesome.

An alternate method of employing the N/2 method is to use a deferred correction approach that we describe below. The N/2 method sequences the elements through 2(N-1) measurements prior to element update. On dither sequence m, m = 1,2,...,N-1, each of the elements n, n = 1,...,N will be shifted $+\pi/2$ for one measurement and $-\pi/2$ for the next measurement, if the Walsh function $W_m[(2n-N-1)/2N] = +1$. Otherwise, the element is not shifted (Fig. 6 with N = 4).

*The initial choice need not be random; it is better to group elements whose phase is better correlated (through proximity) as this should speed convergence.

Let us define $a_{m,n} \triangleq 1/2\ W_m[(2n-N-1)/2N]$, where $a_{m,n} = +1/2$ or $-1/2$. The measured phase deviation $\hat{\phi}_m$, described below, is used to compensate each of the elements n. To describe the by-halves estimation procedure, assume each element has a constant electric field strength E_o and phasor θ_n, for elements $n = 1,2,\ldots,N$. The undithered intensity seen at the detector is

$$I = \left| E_o \sum_{n=1}^{N} e^{i\theta_n} \right| = |E_o|^2 \left[N + 2 \sum_{p=1}^{N} \sum_{q=p+1}^{N} \cos(\theta_p - \theta_q) \right] \tag{8}$$

For small phase angles θ_p, the dither by $\pm\pi/2$ yields the intensities

$$I_m(\pm\pi/2) = N\ |E_o|^2 \left[\frac{N}{2} \mp \frac{2}{N} \left\langle \sum_{n=1}^{N} a_{m,n}\ \theta_n \right\rangle \right] \tag{9}$$

Consequently, a good phase estimator is

$$\hat{\phi}_m = \frac{I_m(-\pi/2) - I_m(\pi/2)}{I_m(-\pi/2) + I_m(\pi/2)} = 2 \sum_{n=1}^{N} (2a_{m,n})\ \theta_n \qquad \text{where } 2a_{m,n} = \pm 1 \tag{10}$$

Using the properties of Walsh functions, it can be shown that

$$\theta_n = \sum_{m=1}^{N} a_{m,n}\ \hat{\phi}_m \tag{11}$$

We can arbitrarily set $\phi_i = 0$, e.g., $\phi_1 = 0$. By matrix manipulation, we obtain an expression that does not use the first Walsh function. Hence

$$\theta_n = \sum_{m=2}^{N} (a_{m,n} - a_{m,1})\ \hat{\phi}_m, \qquad n = 2, 3, \ldots, N \tag{12}$$

Figure 6 illustrates the Walsh functions and $a_{m,n}$ for N = 4. For this particular example, the corresponding matrix $\underline{\theta} = A\underline{\hat{\phi}}$ for N = 4 is given by

$$\begin{vmatrix} \theta_2 \\ \theta_3 \\ \theta_4 \end{vmatrix} = \begin{vmatrix} 0 & 1 & -1 \\ 1 & 1 & 0 \\ 1 & 0 & -1 \end{vmatrix} \cdot \begin{vmatrix} \hat{\phi}_2 \\ \hat{\phi}_3 \\ \hat{\phi}_4 \end{vmatrix} \tag{13}$$

Figures 4 and 5 illustrate computer simulation convergence runs (ratio of actual detector currents to correctly figure currents) for correct-as-you-go systems. Unfortunately, these were run with arc-tan estimators (before the S/N advantage of the arc-sine estimator was appreciated) and therefore are not representatives of best attainable performance.

References

1. V. S. Zuev, E. P. Orlov, and V. A. Sautkin, "Analysis of the possibility of interferometric laser control of the surface of an optical telescope mirror. Part I," Sov. J. Quant. Electron. 5, 43 (1975).

2. K. E. Erickson, "Holographic method of monitoring the performance of a large telescope mirror," Proceedings of Optical Telescope Technology Workshop, Marshall Space Flight Center, April to May 1, 1969, NASA SE-233, 311.

3. W. N. Peters, R. A. Arnold, and S. Gowrinathan, "Stellar interferometer for figure sensing of orbiting astonomical telescopes," Applied Optics 13, 1785 (1974).

4. R. R. Berggren and G. E. Lenertz, "Feasibility of a 30-meter space based laser transmitter," NASA CR-134903, October 1975.

5. W. E. Howell, "Recent advances in optical control for large space telescopes," Space Optics - Proceedings of the 9th International Congress of the International Commission for Optics, National Academy of Sciences, 239 (1973).

6. W. B. Bridges, S. Hansen, L. A. Horwitz, S. P. Lazzara, T. R. O'Meara, J. E. Pearson, and T. J. Walsh, J. Opt. Soc. Am. 64, 541 (1975).

7. J. Hardy, J. Feinleib, and J. C. Wyant, Paper ThBl, OSA Topical Meeting on Optical Propagation through Turbulence, Boulder, Colo., July 1974 (conference proceedings).

8. L. Miller, W. P. Brown, Jr., J. A. Jenney, T. R. O'Meara, Paper ThB2, OSA Topical Meeting on Optical Propagation through Turbulence, Boulder, Colo., July 1974 (conference proceedings).

9. R. A. Muller and A. Buffington, J. Opt. Soc. Am. 64, 1200 (1974).

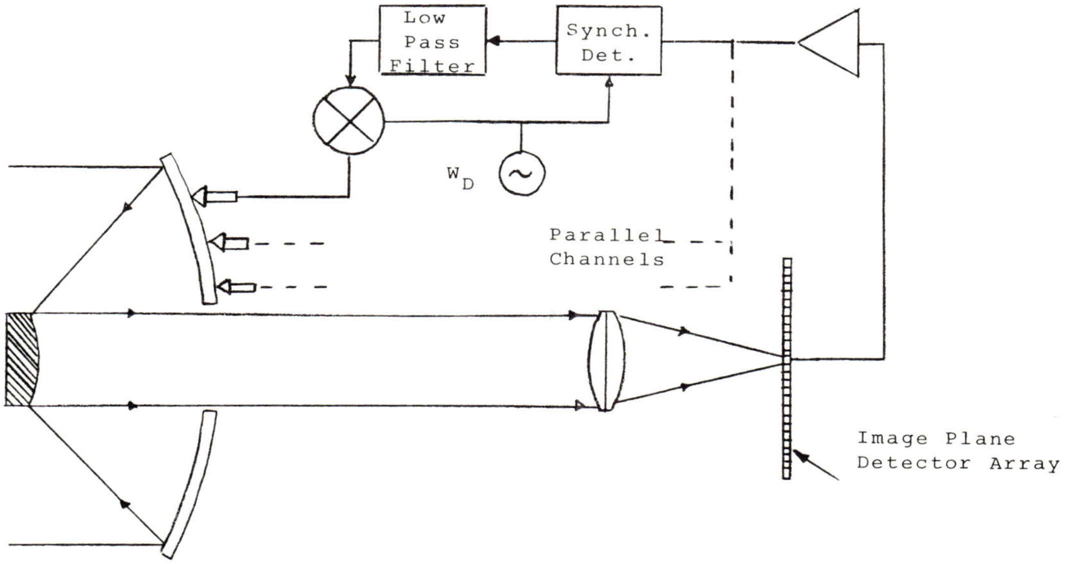

Figure 1. One Class of Imaging-COAT System. An image of an unresolved star is centered on a single (reference) detector in the image plane detector array by control systems (which are not illustrated).

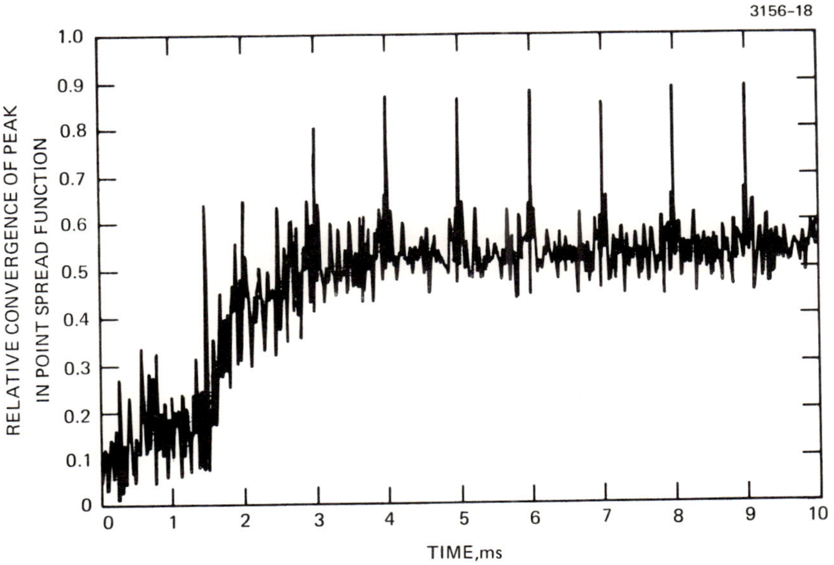

Figure 2. Computer Simulation of Convergence for Sinusoidal Multidither Imaging COAT, ψ_0 = 60°, at 80 photoelectrons/5 μs.

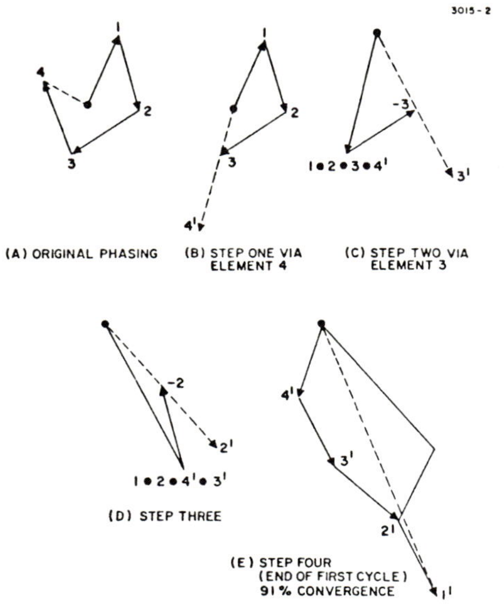

Figure 3(a). Polystep One-Element-at-a-Time Convergence Sequence.

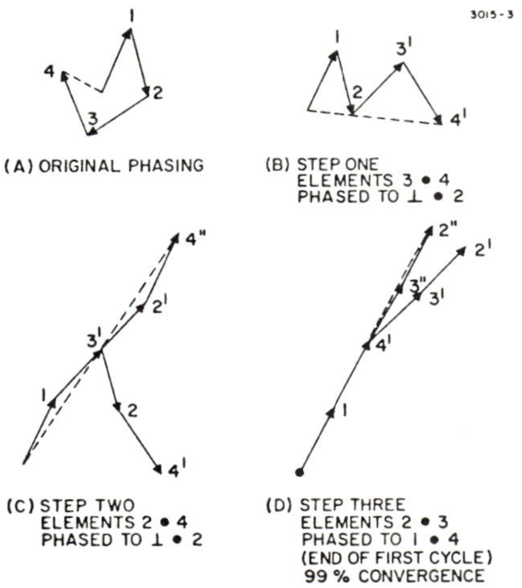

Figure 3(b). Polystep N/2 Elements-at-a-Time Convergence Sequence.

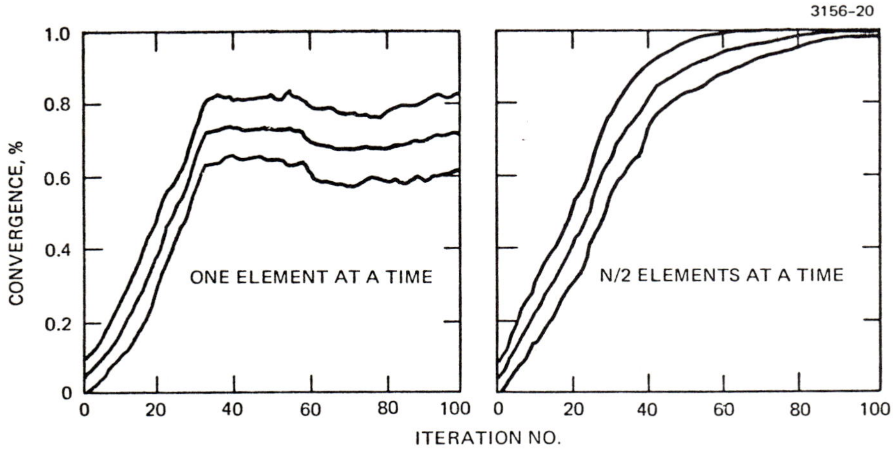

Figure 4. Polystep Convergence Comparison, 10^3 Photoelectrons per Step, 32 Elements.

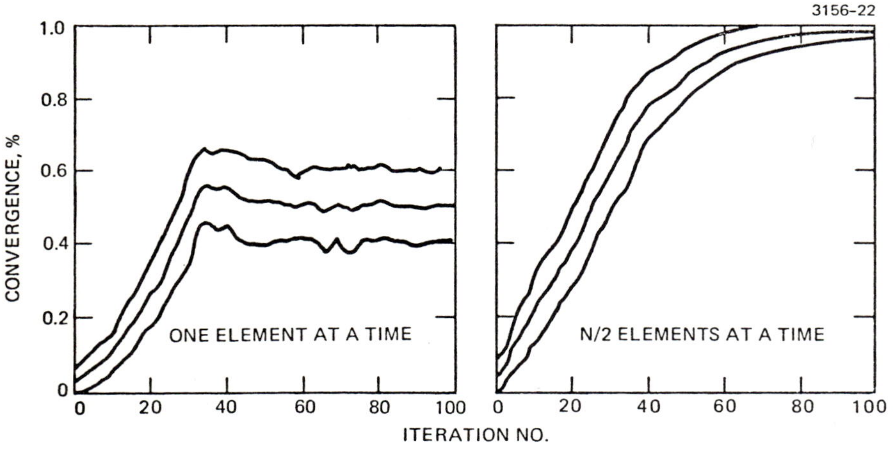

Figure 5. Polystep Convergence Comparison, 500 Photoelectrons per Step, 32 Elements. The central curve is the mean, while the outer curves denote the variance bounds.

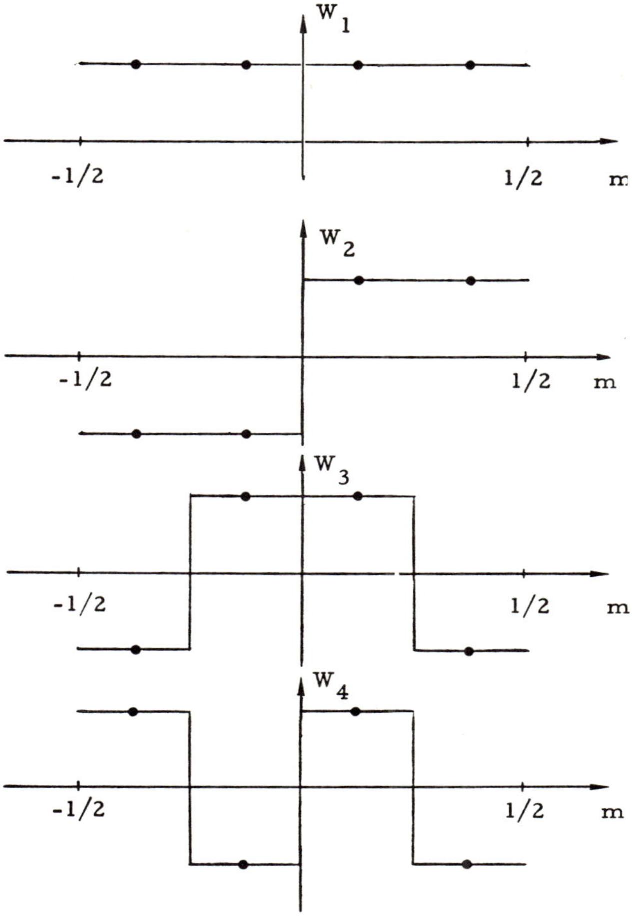

Figure 6. Walsh Functions (1, 2, 3, 4).

Session 5
POST-DETECTION COMPENSATION

Session Chairman
D. L. Fried
Optical Science Consultants

POST-PROCESSING OF IMAGERY FROM ACTIVE OPTICS--SOME PITFALLS

Richard E. Wagner
Optical Sciences Center, University of Arizona
Tucson, Arizona 85721

Abstract

There may be pitfalls to watch out for when recording imagery from active phase compensation devices located in the exit pupil of the system. At least three of these pitfalls are identified and each is manifested as a degradation in the imagery. Both nearby and high altitude atmospheric disturbances contribute to the degradation. The first pitfall is identified as residual phase errors caused by measurement and hardware limitations. The second pitfall is identified as non-isoplanatism and it occurs because wavefronts from different source points experience different high altitude disturbances. The third pitfall is identified as amplitude fluctuations in the exit pupil of the system and it is also caused by the high altitude atmospheric disturbances. The two pitfalls caused by the high altitude disturbances are reduced by applying additional phase compensation in a plane that is the image of the high altitude disturbances. But even this may not be sufficient to eliminate the need for post-processing of the imagery of extended sources.

Introduction

Recently there has been considerable interest in active phase compensation devices that are used to reduce the degrading effects of atmospheric turbulence when imaging sources that lie outside the Earth's atmosphere. Although these devices are expected to go a long way in improving the overall resolution and operation of imaging systems of this type, there are at least three pitfalls that can occur when the device is located in the exit pupil. These pitfalls stem from the inadequacies of the device to perfectly correct for the temporal and spatial variations in optical path length between the source and the image.

To identify these pitfalls it is instructive to consider the degradation as two independent phase disturbances located at different heights in the atmosphere. Although this is not a completely accurate description of the atmosphere, there is good evidence that it is a suitable approximation for many applications[1] and it provides some insight into the problems associated with active phase compensation. A sketch that illustrates the salient points of this model of the atmosphere is shown in Fig. 1.

Atmospheric disturbances occuring far from the optics and disturbances occuring near the optics both affect the wavefront in the exit pupil of the system. In the absence of nearby disturbances the wavefront in the exit pupil has both phase ϕ_h and amplitude ψ modulations caused by the high altitude disturbances. These perturbations are designated by $\exp[\psi(\bar{x}) + i\phi_h(\bar{x})]$ where ψ and ϕ_h are normal random processes described in the exit pupil. Nearby atmospheric disturbances cause only phase perturbations in the exit pupil and these add to the phase perturbations produced by the high altitude disturbances. If the perturbations caused by nearby disturbances are described by $\exp[i\phi_0(\bar{x})]$ then the total wavefront perturbation is given by $\exp[\psi(\bar{x}) + i[\phi_0(\bar{x}) + \phi_h(\bar{x})]]$. Each of the functions ψ, ϕ_h, and ϕ_0 is a random process described in the exit pupil of the system, and each is assumed to be statistically independent of the others.

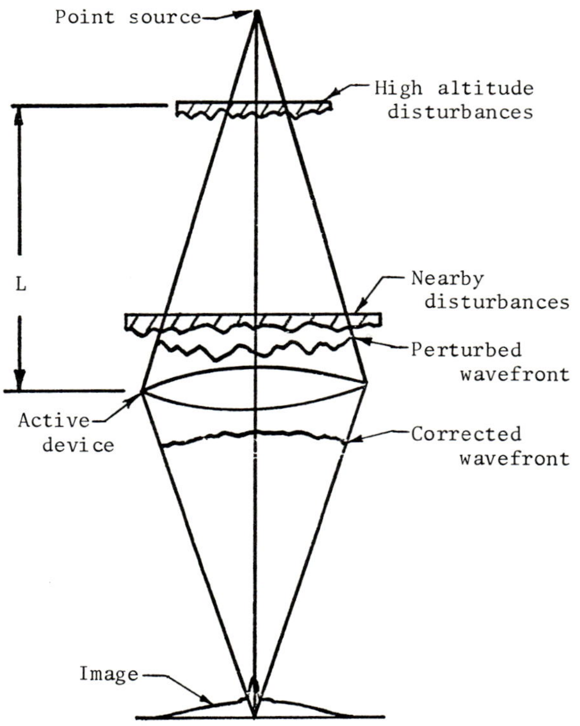

Fig. 1. A simplified model of the atmospheric disturbances with both nearby and high altitude phase perturbations.

The wave structure function defined by Eq. (1) is of primary interest in describing the effects of the atmosphere on the imagery from the optical system.[2]

$$
\begin{aligned}
D(\bar{\tau}) &= D_0(\bar{\tau}) + D_h(\bar{\tau}) + D_\psi(\bar{\tau}) \\
\text{where } D_0(\bar{\tau}) &= E\left\{\left[\phi_0(\bar{x}) - \phi_0(\bar{x} + \bar{\tau})\right]^2\right\} \\
D_h(\bar{\tau}) &= E\left\{\left[\phi_h(\bar{x}) - \phi_h(\bar{x} + \bar{\tau})\right]^2\right\} \\
D_\psi(\bar{\tau}) &= E\left\{\left[\psi(\bar{x}) - \psi(\bar{x} + \bar{\tau})\right]^2\right\}
\end{aligned}
\quad (1)
$$

The wave structure function consists of three terms each of which ultimately leads to a pitfall in the imagery from active compensation devices located in the exit pupil. The first term is the structure function associated with atmospheric disturbances located near the optics; the second term is the structure function associated with phase perturbations in the exit pupil caused by high altitude atmospheric disturbances; the third term is the structure function associated with amplitude variations in the exit pupil caused by atmospheric disturbances far from the optics. The pitfalls corresponding to these three terms are residual phase errors (corresponding to D_0), non-isoplanatism (corresponding to D_h), and amplitude variations (corresponding to D_ψ). Each of these pitfalls is manifested as a degradation in the imagery from active compensation devices, and the effects of each pitfall are discussed below.

Residual Phase Errors

Consider only the phase perturbations caused by nearby atmospheric disturbances. Because of difficult fundamental and practical problems, active compensation devices are not capable of perfectly correcting for these perturbations. Fundamentally the source radiance limits the accuracy with which the phase perturbations are measured;(3) practically the mechanical hardware limits the spatial resolution of the applied correction.(4) The result is that residual errors exist in the wavefront propagating from the active compensation device.

These residual errors degrade the resulting imagery by causing a fraction of the incident power to be scattered into a broad background pattern. On the long time average these residual errors cause the image of a point source to consist of the sum of two decidedly different components.(5) One of these components represents the fraction of power that is not scattered by the phase perturbations and that is identical in form to the image that results when no perturbations are present. It is referred to as the signal of the image. The other component represents the power that is scattered through various angles by the random phase errors and it is called the background. For phase errors that are normally distributed, the fraction of power f in the signal is determined by the rms phase error σ_W.(6)

$$f = e^{-\sigma_W^2} \tag{2}$$

For moderate rms phase errors the width of the background pattern is governed by the correlation width of the residual phase errors. A sketch of the image of a point source illustrating the signal and the background is shown in Fig. 2a for moderate phase errors.

The background causes significant degradation in the imagery even for relatively small rms phase errors. This problem is compounded for imaging extended sources because the background irradiance is proportional to the source area but the signal irradiance is not. For example, two point sources separated by a small distance have background patterns which overlap, but signal patterns which do not. In this case the background irradiance is increased by a factor of two, but the signal irradiance is not. For an extended source with sinusoidal radiance variations, the signal and background are illustrated in Fig. 2b.

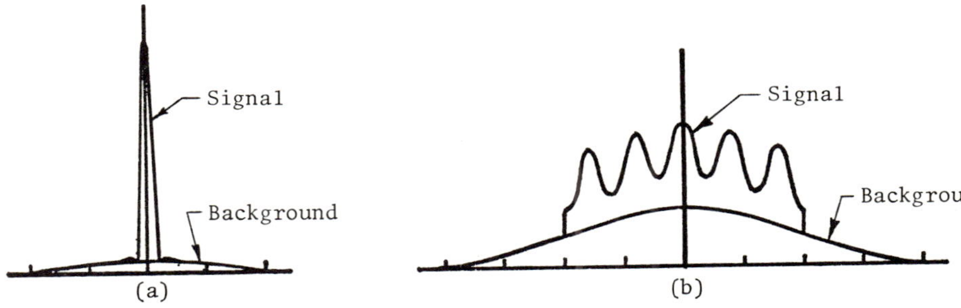

Fig. 2. Image degradation caused by the pitfall of residual phase errors. For $(\sigma_r/2\pi)$ = 1/6 wave rms. (a) Image of a point source; (b) image of an extended source with sinusoidal radiance variations.

The ratio of signal to background irradiance of the image determines the detector SNR required to isolate the signal from the background. It depends strongly upon rms phase error, and upon the source size and spatial frequency content. The SNR required to isolate the signal is shown in Fig. 3a for a source of finite area that has sinusoidal radiance variations. The SNR requirements depend upon source area as shown in the figure, and they are computed for a source spatial frequency for which the optical system has a transfer factor of 0.5.(6) The area of the source referred to the image plane is A_S and the area of the background pattern is A_B. The SNR requirements apply to the image recorded with infinite exposure time.

When the exposure time is finite the spatial fluctuations in the background serve to mask the signal. These fluctuations control the exposure time required to insure that spatial variations in the source are not confused with irradiance fluctuations in the background. The required exposure time T depends strongly upon rms phase error and upon source size and spatial frequency content.(6) For the extended source discussed in the preceeding paragraph the dependence of required exposure time upon source size is illustrated in Fig. 3b. In the figure, τ_c is the temporal correlation time of the irradiance fluctuations of the image.

Non-Isoplanatism

Consider phase perturbations in the exit pupil that are caused by high altitude atmospheric disturbances. Unlike the phase perturbations caused by nearby atmospheric disturbances, these perturbations are field dependent. A wavefront propagating from one source point experiences a different high altitude dis-

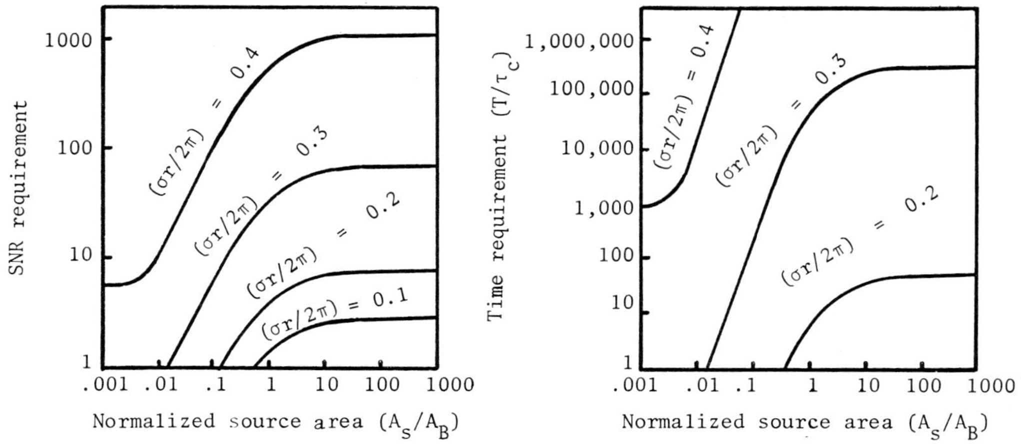

(a) (b)

Fig. 3. Detector requirements for isolation of the signal from the background degradation caused by residual phase errors. The requirements are for an extended source with sinusoidal radiance variations that have a transfer factor of 0.5. (a) Detector SNR requirement; (b) Exposure time requirement.

turbance than a wavefront propagating from another source point. This situation is illustrated in Fig. 4 where the disturbance is located a distance L from the exit pupil and the two sources have an angular separation $\bar{\alpha}$. The wavefront in the exit pupil from the one source is simply shifted by an amount $\bar{\alpha}L$ from the wavefront from the other source. An active compensation device located in the exit pupil cannot compensate for both wavefronts simultaneously. If the device corrects for one of the wavefronts, the ability of it to correct for the other wavefront depends upon the correlation between the two wavefronts.

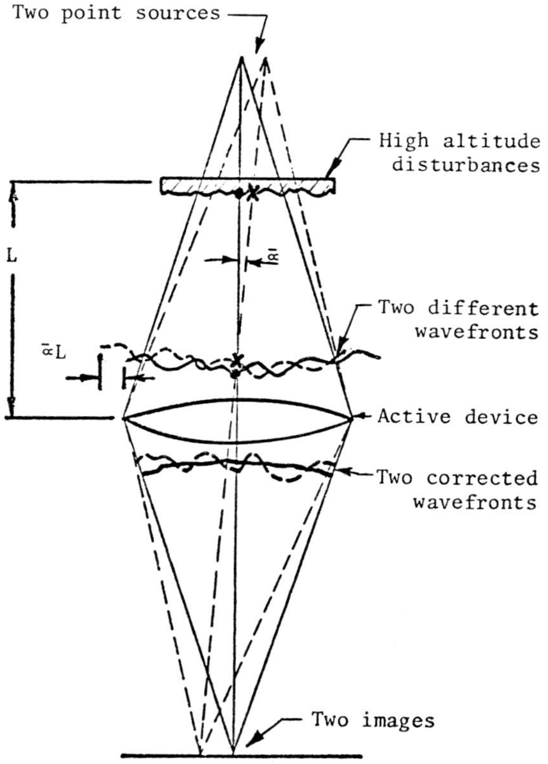

Fig. 4. Non-isoplanatism caused by high altitude atmospheric disturbances. Wavefronts from two different sources are shifted by an amount $\bar{\alpha}L$ in the pupil of the optical system.

Although Korff[7] has discussed the problem of isoplanatism in great detail and with rigor, a much more simplified description of the isoplanatism is fruitful as well. Since the image characteristics are mainly controlled by the rms phase error σ_w, it is derived as a function of field position assuming that the active device corrects for an on-axis point. To do this, let $\phi_c(\bar{x})$ be the phase correction introduced by the active device at a particular time and let $\phi_w(\bar{x},\alpha)$ be the phase perturbations of the corrected wavefront for a source point located at field position $\bar{\alpha}$.

$$\phi_w(\bar{x},\alpha) = \phi_0(\bar{x}) + \phi_h(\bar{x} + \bar{\alpha}L) - \phi_c(\bar{x}) \qquad (3)$$

In Eq. (3) the phase perturbations ϕ_h caused by high altitude atmospheric disturbances depend upon field angle because the wavefront at the exit pupil is shifted by an amount $\bar{\alpha}L$.

The rms phase error of the resulting wavefront is given by the ensemble average of $\phi_w^2(\bar{x},\bar{\alpha})$.

$$\begin{aligned}\sigma_w^2(\bar{\alpha}) &= E\{\phi_w^2(\bar{x},\bar{\alpha})\} \\ &= E\{[\phi_0(\bar{x}) + \phi_h(\bar{x} + \bar{\alpha}L) - \phi_c(\bar{x})]^2\}\end{aligned} \qquad (4)$$

The rms phase error corresponding to $\bar{\alpha} = 0$ is simply the residual phase error σ_r discussed earlier.

$$\sigma_w^2(0) = \sigma_r^2 = E\{[\phi_0(\bar{x}) + \phi_h(\bar{x}) - \phi_c(\bar{x})]^2\} \qquad (5)$$

The rms phase error σ_w described in Eq. (4) is re-written by adding and subtracting the term $\phi_h(\bar{x})$ inside the square brackets and expanding the square. After collecting the terms of the expansion, σ_w is given as a function of field angle.

$$\begin{aligned}\sigma_w^2(\bar{\alpha}) = E\{&[\phi_0(\bar{x}) + \phi_h(\bar{x}) - \phi_c(\bar{x})]^2 \\ &+ [\phi_h(\bar{x} + \bar{\alpha}L) - \phi_h(\bar{x})]^2 \\ &+ 2[\phi_0(\bar{x}) + \phi_h(\bar{x}) - \phi_c(\bar{x})][\phi_h(\bar{x} + \bar{\alpha}L) - \phi_h(\bar{x})]\}\end{aligned} \qquad (6)$$

In Eq. (6) the first term inside the ensemble average is due to the residual errors σ_r defined by Eq. (5). The second term is the structure function $D_h(\bar{\alpha}L)$ defined by Eq. (1). The third term is the correlation between the residual phase errors and the difference in phase perturbations caused by a change in field position. Since these two perturbations arise from completely different phenomena, they are statistically independent and the third term is zero.

After compensation, the wavefront errors as a function of field angle have an rms phase error σ_w given by Eq. (7).

$$\sigma_w^2(\bar{\alpha}) = \sigma_r^2 + D_h(\bar{\alpha}L) \tag{7}$$

Eq. (7) describes the rms phase error for phase perturbations caused by both nearby and high altitude atmospheric disturbances. For $\bar{\alpha} = 0$ the rms phase error is simply the residual rms phase error caused by measurement and hardware limitations. At other field positions the rms phase error after correction increases according to the structure function of the phase perturbations caused by high altitude atmospheric disturbances. According to Eq. (7), the rms phase error σ_w is a strong function of field position and it controls the fraction of power in the signal. The dependence of this fraction on field position is found by combining Eq. (2) and (7).

$$f(\bar{\alpha}) = e^{-\sigma_r^2} \times e^{-D_h(\bar{\alpha}L)} \tag{8}$$

Eq. (8) allows for a practical definition of the size of the isoplanatic region as the maximum field position $\bar{\alpha}_0$ for which the fraction $f(\bar{\alpha}_0)$ is greater than some detectable threshold.

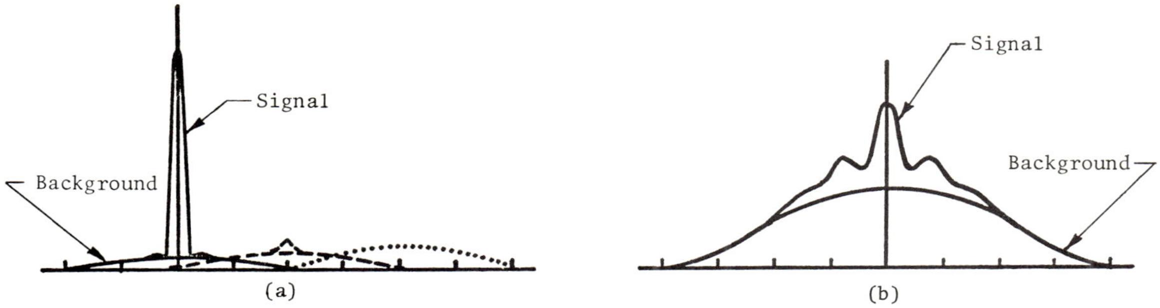

Fig. 5. Image degradation caused by the two pitfalls of residual phase errors and non-isoplanatism. For $(\sigma_r/2\pi) = 1/6$ wave rms and isoplanatic region = background size. (a) Image of three point sources; (b) Image of an extended source with sinusoidal radiance variations.

The pitfall associated with non-isoplanatism is illustrated in Fig. 5a for three point sources. When the active device compensates for the on-axis point source the other two point sources do not have well-corrected images. Power from the background patterns of the two poorly-corrected sources degrades the image of the well-corrected source. This pitfall occurs whenever the background pattern is wider than the isoplanatic region and the severity of the problem depends upon the size of the isoplanatic region compared to the size of the background pattern. In Fig. 5b the combined degradations in the imagery due to residual phase errors and non-isoplanatism are illustrated for the extended source with sinusoidal radiance variations that is discussed earlier. The signal is negligible at field positions where the structure function $D_h(\bar{\alpha}L)$ is large; the well-corrected region of the image is degraded by spillover from the background pattern of poorly-corrected regions of the image.

Amplitude Variations

For amplitude variations in the exit pupil, the fraction of power in the signal is reduced by the amount $\exp[-\frac{1}{2}D\psi(\infty)]$ at all field positions. To get a feeling for the potential of this pitfall, consider the relationship of $D\psi(\infty)$ to the irradiance fluctuations[8] in the exit pupil caused by the high altitude disturbances.

$$\tfrac{1}{2}D\psi(\infty) = \tfrac{1}{4}\ln(1 + \sigma_I^2) \tag{9}$$

In Eq. (9) the variance of the irradiance fluctuations σ_I^2 is normalized by the square of the average irradiance in the exit pupil. By observing the star Arcturus near the zenith, Burke[9] has measured the irradiance fluctuations caused by atmospheric disturbances using a 38mm aperture. These measurements, when corrected for the effects of aperture averaging in strong turbulence[8], indicate that $\sigma_I \simeq 1.3$.

For this level of irradiance fluctuations, the quantity $\frac{1}{2}D\psi(\infty) = \frac{1}{4}\ln[1 + (1.3)^2]$, so that the reduction of power in the signal is given by Eq. (10).

$$e^{-\frac{1}{2}D\psi(\infty)} = \left[\frac{1}{1 + (1.3)^2}\right]^{\frac{1}{4}} = 0.78 \tag{10}$$

In this case amplitude variations reduce the signal by a factor of only three-quarters. Lawrence and Strohbehn[10] indicate that for strong turbulence the irradiance fluctuations saturate, with a maximum value of $\sigma_I = 1.3$ for a plane wave. Even for strong turbulence, the effect of amplitude variations on the image is not a serious problem.

These results are derived for an active device that is located in the exit pupil of the system and they do not apply to devices located at other positions in the system. The two pitfalls of non-isoplanatism and amplitude variations may be reduced by additional phase correction in a plane which is an image of the high altitude disturbance.[11] This technique appears to be essential to produce good image quality for extended sources when high altitude atmospheric disturbances are a problem. Even for systems that provide active phase correction in several planes the two pitfalls of non-isoplanatism and amplitude variations probably will never be completely eliminated, especially when the high altitude disturbance is not localized in one plane. Such techniques enlarge the isoplanatic region and reduce the effects of amplitude variations, but they undoubtedly have residual effects that need to be investigated.

Summary

The imagery from active compensation devices located in the exit pupil contains at least three pitfalls that require careful consideration. For extended sources residual phase errors alone cause a degradation that may require post-processing of the imagery. The degradation is due to a background term that contains a significant fraction of the total power even for small rms phase errors. In addition to residual phase errors, the two pitfalls of non-isoplanatism and amplitude variations further degrade the image. It may be possible to reduce these two problems by applying additional phase correction in a plane that is the image of the high altitude atmospheric disturbance. But to determine the effects of these two pitfalls on the imagery, it is essential to have accurate measurements of the statistics of the perturbations caused by the high altitude atmospheric disturbances alone.

References

1. A. M. Schneiderman, Imaging in Astronomy Technical Digest, paper ThC4, Cambridge, Mass., June 1975.
2. D. L. Fried, J. Opt. Soc. Am. 56, 1372, 1966.
3. J. C. Wyant, Appl. Opt. 14, 2622, 1975.
4. R. H. Hudgin, J. Opt. Soc. Am. 65, 1211, 1975.
5. R. V. Shack, Optical Sciences Center Technical Report 19, University of Arizona, Tucson, Arizona, 1967.
6. R. E. Wagner, J. Opt. Soc. Am. 66, 175, 1976.
7. D. Korff, G. Dryden, and R. J. Leavitt, J. Opt. Soc. Am. 65, 1321, 1975.
8. R. L. Fante, Proc. IEEE 63, 1669, 1975.
9. J. J. Burke, J. Opt. Soc. Am. 60, 1262, 1970.
10. R. S. Lawrence and J. W. Strohbehn, Proc. IEEE 58, 1523, 1970.
11. R. H. Dicke, J. Opt. Soc. Am. 65, 1206, 1975.

FUNDAMENTAL LIMITATIONS IN LINEAR INVARIANT RESTORATION OF ATMOSPHERICALLY DEGRADED IMAGES

J.W. Goodman and J.F. Belsher
Department of Electrical Engineering
Stanford University
Stanford, California 94305

Abstract

Restoration of atmospherically degraded images is limited most fundamentally by the photon noise inherent in any detected image. After presenting a general model for photon-limited images, we derive the form of the linear, space invariant filter which restores the image with minimum mean-squared error. Measures of the restorable bandwidth and image quality are developed. The theory is applied to the case of images degraded by atmospheric turbulence, both with and without perfect tilt removal. The relationship between the number of detected photoevents and the restorability of the degraded images is quantified.

Introduction

The prime practical limitation encountered in attempts to restore degraded imagery arises from noise inherent in the detected image data. The most basic source of such noise lies in the photon fluctuations associated with the detection of the finite amount of light energy available to the imaging system. Thus photon fluctuations pose a fundamental limitation to the "restorability" of a degraded image.

In the first part of this paper we present a model which can be used to mathematically and statistically describe an image detected at low light levels. Following this development, attention is turned to the problem of linear, invariant, least-square restoration of imagery limited by photon noise. The form of the least-square restoration filter is derived using the statistical model appropriate for photon noise. Mathematical measures of the bandwidth and quality of the restored image are developed. Finally, the analytical results are applied to the problem of restoring images degraded by long-time-average atmospheric turbulence.

Modeling of Photon-Limited Imagery

Our model for the detected imagery is essentially a two-dimensional analog of the semi-classical model developed by Mandel[1,2] for the study of photon counting statistics. This semi-classical model is known to yield results that are in complete agreement with a rigorous quantum mechanical model for all detection problems involving the photoelectric effect[3]. Thus the vast majority of image detection problems are properly included in this framework.

Semi-Classical Model

The model utilized is that of an inhomogeneous or compound Poisson impulse process. Thus the detected data $d(x,y)$ is represented by

$$d(x,y) = \sum_{n=1}^{N} \delta(x-x_n, y-y_n) \tag{1}$$

where $\delta(\cdot,\cdot)$ is a two-dimensional Dirac delta function, (x_n,y_n) represents the location of the nth photoevent (i.e., the release of a photoelectron), and N is the total number of photoevents produced by the image. In this representation, N, x_n and y_n are all regarded as random variables, with statistical properties to be described in the following.

In accord with the semi-classical theory of photodetection, the probability that N events occur in an area A on the detector is taken to be Poisson,

$$P_A(N) = \frac{\left[\iint_A \lambda(x,y)\,dx\,dy\right]^N}{N!} \exp\left[-\iint_A \lambda(x,y)\,dx\,dy\right], \tag{2}$$

where the "rate" $\lambda(x,y)$ is related to the classical image intensity $i(x,y)$ falling on the detector through

$$\lambda(x,y) = \frac{\eta\, i(x,y)}{h\bar{\nu}}\,\tau \quad . \tag{3}$$

Here η is the quantum efficiency of the photosurface (assumed independend of (x,y)), h is Planck's constant, $\bar{\nu}$ is the mean optical frequency, and τ is the detector integration time.

Since the distribution $i(x,y)$ of classical intensity is unknown a priori, $\lambda(x,y)$ must ultimately be modeled as a random process. However, in calculating average properties of the detected image, it is often helpful to first treat $\lambda(x,y)$ as a given known function, and then average the results over the statistics of λ. This procedure is entirely consistent with Baye's rule of statistics. We further note for future use that, for a given $\lambda(x,y)$, the "event locations" (x_n,y_n) are independent random variables for different n's, with common probability density function[4]

$$p(x_n,y_n) = \frac{\lambda(x_n,y_n)}{\iint_{-\infty}^{\infty} \lambda(x,y)\,dx\,dy} . \tag{4}$$

Note in addition that, because $\lambda(x,y)$ is proportional to the classical intensity, $\lambda(x,y) \geq 0$.

Before turning to an examination of the image properties implied by this model, we point out some of the ways it can be generalized. First, in practice the photoevents registered by a real distributed detector consist of finite spatial pulses, not delta functions. This fact can be incorporated in our model by passing our ideal detected data (Eq.(1)) through a linear spatial filter, thus spreading the delta functions into pulses. Second, due to photoelectron multiplication noise, the areas and shapes of the various pulses may themselves be random variables. This property can be included in our model by making the spatial filter randomly space-variant. These sophistications can all be included in the model, but since our interest lies in fundamental limits, there is little to be gained by incorporating these additional non-fundamental degradations.

We close this section by showing in Fig. 1 a typical classical intensity distribution and a corresponding typical detected image, the illustration being one-dimensional for simplicity.

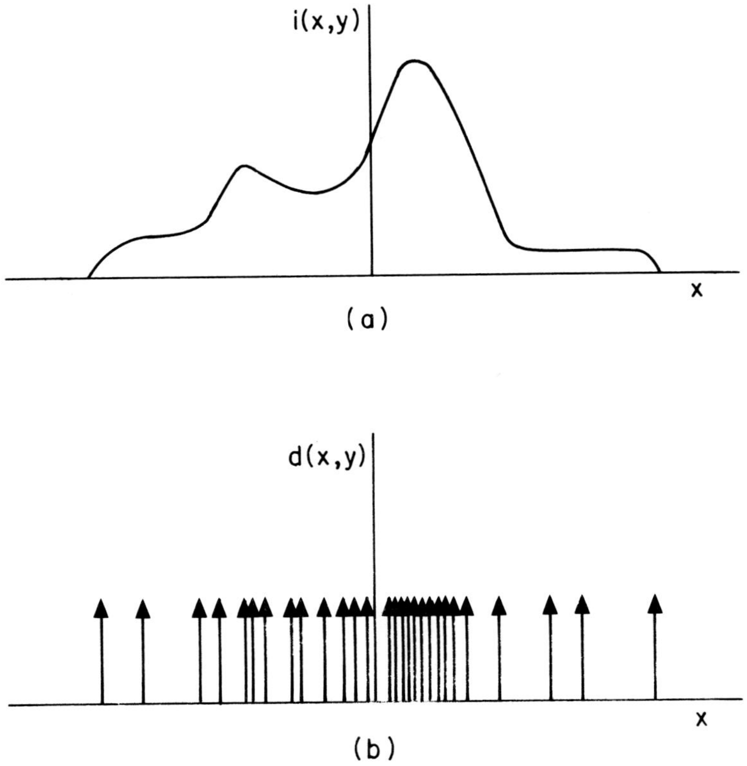

Fig. 1. Model of photon-limited imagery: (a) classical intensity incident on detector; (b) resulting detected image.

Spectral Density of Photon-Limited Imagery

One of the fundamental properties of a detected image is its spectral density. Our goal in this section is to calculate the spectral density of the detected image described by Eq.(1). If $D(\nu_X,\nu_Y)$ represents the Fourier transform of $d(x,y)$, i.e.,

$$D(\nu_X,\nu_Y) = \iint_{-\infty}^{\infty} d(x,y)e^{-j2\pi(\nu_X x+\nu_Y y)}\,dxdy \quad , \tag{5}$$

then our goal is to calculate

$$\Phi_d(\nu_X,\nu_Y) = E\left[|D(\nu_X,\nu_Y)|^2\right] \quad , \tag{6}$$

where the symbol $E[\cdot]$ indicates an expectation over the statistics of N , (x_n,y_n) , and λ. Before making this calculation, we must first mention a physical restriction we impose on the random process $\lambda(x,y)$, i.e., on the statistical properties of the classical intensity in the image plane.

Any image produced by an optical system must contain finite total energy. From this fact it follows that for every sample function of the random process $\lambda(x,y)$,

$$\iint_{-\infty}^{\infty} \lambda(x,y)\,dxdy < \infty \quad . \tag{7}$$

This equation is sufficient to imply that any particular sample function $\lambda(x,y)$ has a well-defined Fourier transform,

$$\Lambda(\nu_X,\nu_Y) = \iint_{-\infty}^{\infty} \lambda(x,y)e^{-j2\pi(\nu_X x+\nu_Y y)}\,dxdy \quad . \tag{8}$$

Our goal now is to relate the spectral density $\Phi_d(\nu_X,\nu_Y)$ of the detected image to the spectral density

$$\Phi_\lambda(\nu_X,\nu_Y) = E\left[|\Lambda(\nu_X,\nu_Y)|^2\right] \tag{9}$$

where the expectation is over the statistics of $\lambda(x,y)$.

To calculate the spectral density $\Phi_d(\nu_X,\nu_Y)$ of the detected image, we first Fourier transform Eq.(1), with the result

$$D(\nu_X,\nu_Y) = \sum_{n=1}^{N} e^{-j2\pi(\nu_X x_n+\nu_Y y_n)} \quad . \tag{10}$$

The squared modulus of this quantity is

$$|D(\nu_X,\nu_Y)|^2 = \sum_{n=1}^{N}\sum_{m=1}^{N} e^{-j2\pi\left[\nu_X(x_n-x_m)+\nu_Y(y_n-y_m)\right]} \quad . \tag{11}$$

It remains to find the expected value of this quantity over the statistics of N , (x_n,y_n) and $\lambda(x,y)$. It is convenient to first regard N and $\lambda(x,y)$ as given (known) quantities, average over the statistics of (x_n,y_n) and (x_m,y_m) , and then average over N and λ. Thus our first goal is to compute

$$E_{nm}\left[|D(\nu_X,\nu_Y)|^2\right] = \sum_{n=1}^{N}\sum_{m=1}^{N} E_{nm}\left[\exp\left\{-j2\pi\left[\nu_X(x_n-x_m)+\nu_Y(y_n-y_m)\right]\right\}\right] \tag{12}$$

where E_{nm} signifies an average over (x_n,y_n) and (x_m,y_m).

Two classes of terms can be identified. First, there are N terms for which $n = m$, each of which yields unity. Second, there are $N^2 - N$ terms for which $n \neq m$. For such terms we know that (x_n,y_n) and (x_m,y_m) are independent random variables, and therefore that

$$p(x_n,y_n;x_m,y_m) = \frac{\lambda(x_n,y_n)}{\iint_{-\infty}^{\infty} \lambda(x,y)dxdy} \cdot \frac{\lambda(x_m,y_m)}{\iint_{-\infty}^{\infty} \lambda(x,y)dxdy} \quad . \tag{13}$$

For these $N^2 - N$ terms, the result of the averaging process is

$$E_{nm}\left[e^{-j2\pi\left[\nu_X(x_n-x_m)+\nu_Y(y_n-y_m)\right]}\right] = \left|\frac{\iint_{-\infty}^{\infty} \lambda(x,y)e^{-j2\pi(\nu_X x+\nu_Y y)dxdy}}{\iint_{-\infty}^{\infty} \lambda(x,y)dxdy}\right|^2 \quad . \tag{14}$$

The result of averaging $|D(\nu_X,\nu_Y)|^2$ over the statistics of (x_n,y_n) and (x_m,y_m) becomes

$$E_{nm}\left[|D(\nu_X,\nu_Y)|^2\right] = N + (N^2-N)\left|\frac{\Lambda(\nu_X,\nu_Y)}{\Lambda(0,0)}\right|^2 \quad . \tag{15}$$

Continuing our averaging process, we next find the expected value of Eq. (15) over the random variable N , given $\lambda(x,y)$. Representing the conditional mean of N (given λ) by $\overline{N}_{(\lambda)}$, and noting that, for Poisson statistics,

$$E\left[N^2-N\right] = \left(\overline{N}_{(\lambda)}\right)^2 \quad ,$$

$$\overline{N}_{(\lambda)} = \iint_{-\infty}^{\infty} \lambda(x,y)dxdy = \Lambda(0,0) \quad , \tag{16}$$

we find that

$$E_{n,m,N}\left[|D(\nu_X,\nu_Y)|^2\right] = \overline{N}_{(\lambda)} + |\Lambda(\nu_X,\nu_Y)|^2 \quad . \tag{17}$$

Finally, averaging over the statistics of $\lambda(x,y)$, we obtain

$$\Phi_d(\nu_X,\nu_Y) = \overline{N} + \Phi_\lambda(\nu_X,\nu_Y) \tag{18}$$

where $\overline{N}$ is the unconditional mean of N.

Thus the spectral density of the detected image is the sum of a constant spectral level $\overline{N}$, plus the spectral density of the rate function, Φ_λ. Alternate forms of this result are also useful. First, if we define a normalized spectral density

$$\hat{\Phi}_\lambda(\nu_X,\nu_Y) = \frac{\Phi_\lambda(\nu_X,\nu_Y)}{\Phi_\lambda(0,0)} \quad , \tag{19}$$

then we have that

$$\Phi_d(\nu_X,\nu_Y) = \overline{N} + (\overline{N})^2\hat{\Phi}_\lambda(\nu_X,\nu_Y) \quad . \tag{20}$$

Furthermore, since $\lambda(x,y)$ is proportional to the classical intensity $i(x,y)$, we must have

$$\hat{\Phi}_\lambda(\nu_X,\nu_Y) = \hat{\Phi}_i(\nu_X,\nu_Y) \tag{21}$$

where $\hat{\Phi}_i(\nu_X,\nu_Y)$ is the spectral density of the classical intensity, normalized to unity at $\nu_X = \nu_Y = 0$. Thus, we write our final result,

$$\Phi_d(\nu_X,\nu_Y) = \overline{N} + (\overline{N})^2\hat{\Phi}_i(\nu_X,\nu_Y) \quad . \tag{22}$$

We illustrate this result in Fig. 2.

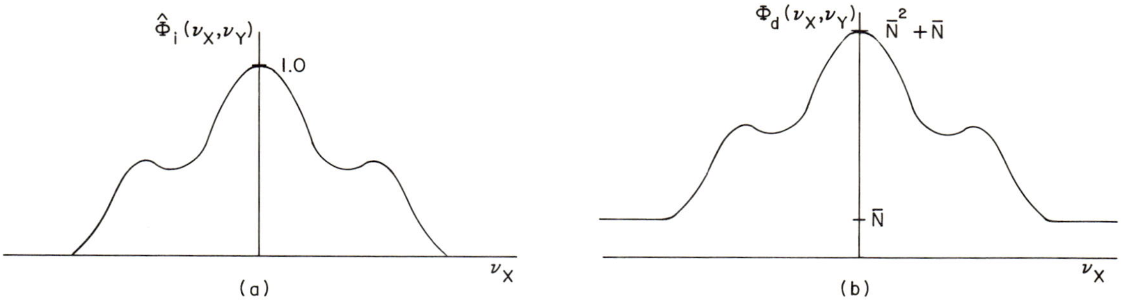

Fig. 2. Relation between (a) normalized spectral density of the classical intensity distribution, and (b) spectral density of the detected image.

Linear Least-Squares Restoration of Degraded Photon-Limited Images

In many practical problems of interest, the detected image data arises from a blurred image of the object of interest. For example, the object may suffer significant motion during the detection interval τ , thus blurring the image. Alternatively, and of greater interest here, the detected image may be seriously degraded by the spatial and temporal fluctuations of the refractive index of the atmosphere, i.e., by "atmospheric seeing". At low light levels, the detected image further suffers from photon noise of the type discussed in the previous section. In order to extract as much information as possible about the object from the detected image data, we seek a method of image restoration which will enhance object detail without unduly emphasizing the noise associated with the discrete photo-events composing the detected image. In the sections to follow, we consider one approach to this problem.

Linear, Invariant Least-Squares Restoration

The approach we shall investigate here is commonly referred to as linear, invariant least-squares restoration. The philosophy behind this approach is best explained with the help of Figure 3.

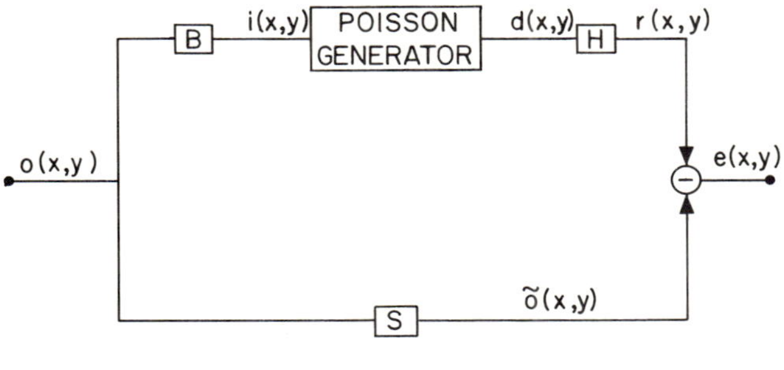

Fig. 3. Block diagram for least-mean-square-error restoration

The function $o(x,y)$ represents the true object brightness distribution, or alternatively the image that would be produced by an ideal optical system (free from aberrations and free from any blur due to diffraction) and an ideal noise-free detector. Our goal is to determine $o(x,y)$ from the actual detected data with the greatest possible accuracy. Following the upper branch of Fig. 3, the ideal object suffers a perfectly known blur on passage to the image plane, this blur being introduced by diffraction, fixed aberrations, and other external causes, such as object motion, atmospheric seeing, etc. We assume that all of these blurs can be lumped together and represented by a single known linear, space-invariant filter, with impulse response $b(x,y)$, or optical transfer function $\hat{B}(\nu_X,\nu_Y)$,

$$\hat{B}(\nu_X,\nu_Y) = \frac{\iint_{-\infty}^{\infty} b(x,y)e^{-j2\pi(\nu_X x+\nu_Y y)}dxdy}{\iint_{-\infty}^{\infty} b(x,y)dxdy} \quad . \tag{23}$$

To represent the statistical fluctuations introduced by the detection process, the blurred image is now applied to a "Poisson generator", which produces a Poisson impulse process with rate $\lambda(x,y)$ proportional to the intensity of the blurred image. The output of the Poisson generator is the detected image data $d(x,y)$, on which we must base our estimate of $o(x,y)$.

Our restoration procedure is to apply the detected image data to a linear, space-invariant restoration filter with impulse response $h(x,y)$ and optical transfer function $\hat{H}(\nu_X,\nu_Y)$. This transfer function will be chosen to minimize the expected value of the mean-squared error,

$$E\left[\iint_{-\infty}^{\infty} e^2(x,y)dxdy\right] = E\left[\iint_{-\infty}^{\infty} \{r(x,y)-\tilde{o}(x,y)\}^2 dxdy\right] \quad , \tag{24}$$

where the error $e(x,y)$ represents the difference between the restored image $r(x,y)$ and a certain filtered version of the object, $\tilde{o}(x,y)$, the expectation being over the statistics of $o(x,y)$ and the statistics of the detection process.

The choice of a filtered object $\tilde{o}(x,y)$, rather than the true object $o(x,y)$, for defining the error requires some comment. The restoration of object frequency components beyond the diffraction-limited cutoff of the optical system is impossible to achieve with any linear invariant restoration filter. Hence the best we can hope to accomplish is restoration of those frequency components lying within the diffraction-limited passband. Accordingly, we count as error only the differences between the restored spectrum $R(\nu_X,\nu_Y)$ and the (possibly modified) portion of the spectrum lying within the observable passband. The frequency spectrum of $\tilde{o}(x,y)$ is thus

$$\tilde{O}(\nu_X,\nu_Y) = \begin{cases} O(\nu_X,\nu_Y)\hat{S}(\nu_X,\nu_Y) & (\nu_X,\nu_Y) \text{ in observable passband} \\ 0 & \text{otherwise} \end{cases} \tag{25}$$

Usually we will take $\hat{S}(\nu_X,\nu_Y)$ to be of the form

$$\hat{S}(\nu_X,\nu_Y) = \begin{cases} 1 & (\nu_X,\nu_Y) \text{ in observable passband} \\ 0 & \text{otherwise} \end{cases} \tag{26}$$

However, we note that the resulting $\tilde{o}(x,y)$ can have negative values in this case. We could alternatively choose $\hat{S}(\nu_X,\nu_Y)$ to be the diffraction-limited optical transfer function of the system, thus guaranteeing a positive $\tilde{o}(x,y)$. In the analysis to follow we leave $\hat{S}(\nu_X,\nu_Y)$ completely general, but eventually we choose the form of Eq.(26) for mathematical simplicity.

Some final comments are in order regarding the non-optimality of the restoration procedure described above. First, it is well known that linear least-squares restoration is not optimal in the sense of maximum likelihood or maximum a posteriori probability when the image statistics are Poisson. Rather, non-linear filtering is required for true optimality[5]. Secondly, the choice of a space-invariant linear restoration filter undoubtedly reduces performance even further; it seems clear intuitively that a space-variant filter can perform better than a space-invariant filter in the presence of signal-dependent noise. However, it should be pointed out that both optimal non-linear filtering and optimal space-variant linear filtering are in general computationally less efficient than linear space-invariant filtering. For this reason, there remains a strong interest in knowing the limitations of linear space-invariant least-squares restoration for photon-limited imagery.

The Form of the Restoration Filter

In this section we derive the form of the linear space-invariant least-square restoration filter for photon-limited images. Our goal is to choose a filter transfer function H which minimizes $E[\iint_{-\infty}^{\infty} e^2(x,y)dxdy]$. By Parseval's theorem, it is equivalent to minimize $E[\iint_{-\infty}^{\infty} |\mathcal{E}(\nu_X,\nu_Y)|^2 d\nu_X d\nu_Y]$, where $\mathcal{E}$ is the Fourier transform of e. Interchanging orders of integration and expectation, we find that it suffices to minimize at each (ν_X,ν_Y)

$$E\left[|\mathcal{E}(\nu_X,\nu_Y)|^2\right] = E\left[|DH - \tilde{O}|^2\right] \tag{27}$$

where $\tilde{O}(\nu_X,\nu_Y)$ is the Fourier transform of $\tilde{o}(x,y)$. The minimization is straightforward and yields

$$H(\nu_X,\nu_Y) = \frac{\Phi^*_{d\tilde{o}}}{\Phi_d} \tag{28}$$

where Φ_d is the spectral density of the detected image $d(x,y)$, while $\Phi_{d\tilde{o}}$ is the cross-spectral density of $d(x,y)$ and $\tilde{o}(x,y)$,

$$\Phi_{d\tilde{o}}(\nu_X,\nu_Y) = E\left[D(\nu_X,\nu_Y)\tilde{O}^*(\nu_X,\nu_Y)\right] \quad . \tag{29}$$

Straightforward calculations, using the Poisson impulse model, show that

$$\Phi_d(\nu_X,\nu_Y) = \overline{N} + (\overline{N})^2|\hat{B}(\nu_X,\nu_Y)|^2\hat{\Phi}_o(\nu_X,\nu_Y)$$

$$\Phi^*_{d\tilde{o}}(\nu_X,\nu_Y) = (\overline{N})^2\hat{S}(\nu_X,\nu_Y)\hat{B}^*(\nu_X,\nu_Y)\hat{\Phi}_o(\nu_X,\nu_Y) \quad . \tag{30}$$

We conclude that the transfer function of the restoration filter is given by

$$H(\nu_X,\nu_Y) = \frac{\overline{N}\,\hat{S}(\nu_X,\nu_Y)\hat{B}^*(\nu_X,\nu_Y)\hat{\Phi}_o(\nu_X,\nu_Y)}{1 + \overline{N}|\hat{B}(\nu_X,\nu_Y)|^2\hat{\Phi}_o(\nu_X,\nu_Y)} \quad . \tag{31}$$

Note that at zero spatial frequency the gain of this filter is $\overline{N}/(1+\overline{N})$. Thus the normalized restoration transfer function is

$$\hat{H}(\nu_X,\nu_Y) = \frac{(1+\overline{N})\hat{S}(\nu_X,\nu_Y)\hat{B}^*(\nu_X,\nu_Y)\hat{\Phi}_o(\nu_X,\nu_Y)}{1 + \overline{N}|\hat{B}(\nu_X,\nu_Y)|^2\hat{\Phi}_o(\nu_X,\nu_Y)} \quad . \tag{32}$$

Quality of the Restored Image

The quality of the restored image is degraded by two separate but related phenomena. First, due to the low signal-to-noise ratios available at high spatial frequencies, the least-square filter wisely attempts to restore only the lower spatial frequencies, where the signal-to-noise ratio is high. Hence, some residual blurring of the restored image remains, thus comprising one component of the mean-square error. In addition, some shot noise is present at the lower spatial frequencies, and passes through the restoration filter, thus contributing a second component of mean-square error.

Recognizing that residual mean-square error will always be present, the question then arises as to how to specify the quality of the restored image. This is an extremely complex subject, for the answer must depend on what is to be done with the restored image once it is obtained. We present two different quality measures, neither of which is entirely satisfactory, but each of which yields some insight into the problem.

One measure of quality which is quite useful is the bandwidth of the product of the blur OTF and the restoration OTF. A suitable measure of this two-dimensional bandwidth is

$$(\Delta\nu)^2 = \iint_{-\infty}^{\infty} |\hat{B}\hat{H}|\,d\nu_X d\nu_Y \tag{33}$$

where

$$\hat{B}\hat{H} = \frac{(1+\overline{N})\hat{S}(\nu_X,\nu_Y)|\hat{B}(\nu_X,\nu_Y)|^2\hat{\Phi}_o(\nu_X,\nu_Y)}{1 + \overline{N}|\hat{B}(\nu_X,\nu_Y)|^2\hat{\Phi}_o(\nu_X,\nu_Y)} \quad . \tag{34}$$

This measure provides an accurate indication of how well the blur has been removed from the picture. It also is dependent on the noise in the detected image, through the dependence of $\hat{H}$ on that noise. However, $(\Delta\nu)^2$ does not, per se, provide a specific indication of the amount of residual shot noise present in the restored image. All we know is the bandwidth that corresponds to an optimum balancing of shot noise and residual blur.

A second possible measure, which accounts for both residual blur and residual shot noise, is based upon the total mean-squared error of the restored image. The total mean-squared error can be expressed as

$$\varepsilon = E\left[\iint_{-\infty}^{\infty} |\mathcal{E}(\nu_X,\nu_Y)|^2 d\nu_X d\nu_Y\right]$$

$$= \iint_{-\infty}^{\infty} \left[|H|^2\Phi_d - H\Phi_{d\tilde{o}} - H^*\Phi^*_{d\tilde{o}} + \Phi_{\tilde{o}}\right] d\nu_X d\nu_Y \quad . \tag{35}$$

Substituting (30) and (31) in (35), we obtain, after some algebra

$$\varepsilon = \iint_{-\infty}^{\infty} \frac{(\overline{N})^2|\hat{S}|^2\hat{\Phi}_o}{1 + \overline{N}|\hat{B}|^2\hat{\Phi}_o} d\nu_X d\nu_Y \quad . \tag{36}$$

The total mean-square error in the image is not a particularly meaningful quantity in itself. Note in particular that, as $\overline{N}$ grows large, so too does ε, in direct proportion to $\overline{N}$. However, the mean-square value of the object within the observable passband is $(\overline{N})^2\iint_{-\infty}^{\infty} |\hat{S}|^2\hat{\Phi}_o d\nu_X d\nu_Y$; hence the "signal" component of the output power rises in proportion to $(\overline{N})^2$, yielding a net increase in restored image quality as $\overline{N}$ increases. As our measure of image quality we choose what might be called a "signal-to-noise ratio", although the term "noise" also includes error due to residual image blur. Noting that as $\overline{N} \to 0$, $\varepsilon \to (\overline{N})^2\iint_{-\infty}^{\infty} |\hat{S}|^2\hat{\Phi}_o d\nu_X d\nu_Y$, we find the proper definition of this quality factor to be

$$Q = \frac{(\overline{N})^2 \iint_{\infty}^{\infty} |\hat{S}|^2\hat{\Phi}_o d\nu_X d\nu_Y}{\varepsilon} - 1 \quad . \tag{37}$$

Alternative equivalent expressions are

$$Q = \overline{N}\,\frac{\displaystyle\iint_{-\infty}^{\infty} \frac{|\hat{S}|^2|\hat{B}|^2\hat{\Phi}_o}{1 + \overline{N}|\hat{B}|^2\hat{\Phi}_o} d\nu_X d\nu_Y}{\displaystyle\iint_{-\infty}^{\infty} \frac{|\hat{S}|^2\hat{\Phi}_o}{1 + \overline{N}|\hat{B}|^2\hat{\Phi}_o} d\nu_X d\nu_Y} \tag{38a}$$

$$= \overline{N}\,\frac{\displaystyle\iint_{-\infty}^{\infty} \frac{|\hat{S}|^2|\hat{B}|^2\hat{\Phi}_o}{1 + \overline{N}|\hat{B}|^2\hat{\Phi}_o} d\nu_X d\nu_Y}{\displaystyle\iint_{-\infty}^{\infty} |\hat{S}|^2\hat{\Phi}_o d\nu_X d\nu_Y - \overline{N}\iint_{\infty}^{\infty} \frac{|\hat{S}|^2|\hat{B}|^2\hat{\Phi}_o}{1 + \overline{N}|\hat{B}|^2\hat{\Phi}_o} d\nu_X d\nu_Y} \tag{38b}$$

Note that for large $\overline{N}$, this quality measure increases linearly with $\overline{N}$.

In future discussions of restored image quality, we present results both in terms of the quality factor Q and in terms of the "maximum restorable frequency" $\Delta\nu/\sqrt{\pi}$.

Dependence on Object Complexity

The results of the previous sections have demonstrated that the degree to which restoration is possible depends not only on the total light flux $(\overline{N})$ and the optical transfer function of the blur $(\hat{B})$, but also on the normalized spectral density of the object $(\hat{\Phi}_o)$. In this section we explore the dependence of this normalized spectrum on "object complexity". To explore this question in a completely general way is extremely difficult. Accordingly, we examine two specific models of the object, neither of which is entirely realistic, but from which we can deduce trends valid for more general object models.

For comparison purposes we note that the simplest possible object is an ideal point source, with brightness distribution

$$o(x,y) = o_o\delta(x-x_1,y-y_1) \quad . \tag{39}$$

While such an object cannot exist physically, nonetheless it serves as a useful idealized case against which we can compare the results of complicated object models. For the point source described above, the normalized spectral density is given by

$$\hat{\Phi}_o(\nu_X,\nu_Y) = 1 \qquad \text{(all } \nu_X,\nu_Y) \quad . \tag{40}$$

The first model utilized to represent a more complicated object is a natural generalization of the case of a single point source. We suppose that there exist M equally intense point sources,

$$O(x,y) = \sum_{m=1}^{M} o_o \delta(x-x_m, y-y_m) \qquad . \qquad (41)$$

We suppose that the locations (x_m, y_m) are independent random variables, uniformly distributed over a square object field of size $L\times L$. Omitting the calculations, which are straightforward, we find the normalized object spectral density in this case to be

$$\hat{\Phi}_o(\nu_X,\nu_Y) = \frac{1}{M} + \left(1 - \frac{1}{M}\right) \mathrm{sinc}^2 L\nu_X \mathrm{sinc}^2 L\nu_Y \qquad (42)$$

which is shown in Fig. 4a. Note that an increase of object complexity has led to a decrease in the level of the normalized spectral density, except at extremely low spatial frequencies.

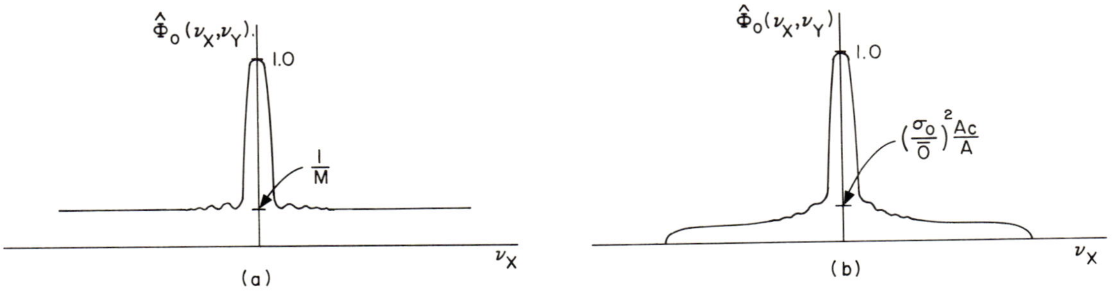

Fig. 4. Normalized spectral densities for two object models: (a) M independent point sources, and (b) stationary object over a finite region.

What is the effect of object complexity, represented by M, on the quality of the restored image? Using the fact that $\hat{\Phi}_o \cong 1/M$ over most spatial frequencies of interest, examination of Eqs. (38a,b) shows that Q is primarily a function of the total number of photoevents contributed by a single point source,

$$\bar{n} = \frac{\bar{N}}{M} \qquad . \qquad (43)$$

If $\bar{n}$ is held fixed, Q is essentially independent of how many point sources are in the field.

A second and somewhat more general model for the object can be formulated as follows. Let the object be modeled as a stationary random process, $o(x,y) \geq 0$, with mean $\bar{o} \geq 0$ and variance σ_o^2. We further suppose that this stationary random process is confined to an $L\times L$ square and is zero outside that square. This space limitation can be explicitly introduced by multiplying the stationary $o(x,y)$ by a window function $\mathrm{rect}(x/L)\mathrm{rect}(y/L)$. We wish to calculate the normalized spectral density for this object model. The calculation is tedious but again straightforward. To state the results in succinct form, we define the following additional symbols:

$A = L^2$ represents the area of the object field;

A_c represents the correlation area of the random process $o(x,y)$, and is specifically defined as

$$A_c = \iint_{-\infty}^{\infty} \gamma_o(\Delta x, \Delta y)\, d\Delta x\, d\Delta y$$

where γ_o is the autocovariance of $o(x,y)$, normalized to unity at the origin; and

$\hat{\Phi}_{\delta o}(\nu_X,\nu_Y)$ is the spectral density of the fluctuations of the object about its mean, normalized to unity at the origin.

With these definitions, the normalized spectral density can be expressed approximately as

$$\hat{\Phi}_o(\nu_X,\nu_Y) \cong \left[1 - \left(\frac{\sigma_o}{\bar{o}}\right)^2 \frac{A_c}{A}\right] \mathrm{sinc}^2 L\nu_X \; \mathrm{sinc}^2 L\nu_Y + \left(\frac{\sigma_o}{\bar{o}}\right)^2 \frac{A_c}{A}\, \hat{\Phi}_{\delta o}(\nu_X,\nu_Y) \tag{44}$$

where the chief approximation is that

$$\left(\frac{\sigma_o}{\bar{o}}\right)^2 \frac{A_c}{A} << 1 \quad . \tag{45}$$

Figure 4b shows a typical plot of this normalized spectral density. Of most importance, we note that, except at the lowest spatial frequencies, the normalized spectral density has a value less than $(\sigma_o/\bar{o})^2(A_c/A)$. This parameter plays a role similar to $(M)^{-1}$ in the previous model. In this case we define the parameter $\bar{n}$ as the mean number of photoevents contributed by a single correlation area of the object

$$\bar{n} = \frac{\bar{N}}{A} \cdot A_c \quad . \tag{46}$$

Again referring to Eqs. (38a,b), we find that the quality of the restored image will depend primarily on the parameter $\bar{n}(\sigma_o/\bar{o})^2$ for any given blur.

Finally, we emphasize that, for both models, the quality of the restored image depends on the number of photoevents generated by a single degree of freedom of the true object, and not on the number of photoevents arising from a single diffraction-limited resolution element, as it often assumed.

Application to Atmospherically Degraded Images

Our goal here is to apply the previous results to the specific case of atmospherically degraded images. All imagery considered here will be assumed to be recorded with an exposure time that is much longer than the characteristic fluctuation time of the atmosphere. Thus we are dealing only with long-exposure imagery. However, we consider both ordinary long-exposure imagery and "tilt-removed" long-exposure imagery. In the second case it is assumed that a perfect tilt removal system operates to keep the image perfectly centered on a fixed point at all times.

Using the results of Fried[6], we have that the OTF of atmospherically induced blur is given by

$$\hat{B}_A(\nu) = \exp\left\{-3.44\left(\frac{\bar{\lambda}\nu F}{r_o}\right)^{5/3}\left[1 - \alpha\left(\frac{\bar{\lambda}\nu F}{D}\right)^{1/3}\right]\right\} \tag{47}$$

where

$\bar{\lambda}$ is the mean wavelength;

F is the focal length of the telescope;

D represents the telescope diameter;

$\nu = \sqrt{\nu_X^2 + \nu_Y^2}$ represents radius in the spatial frequency plane;

r_o represents the coherence parameter of the atmospheric wavefront distortions, as defined by Fried[6] ; and

α takes on the value zero, one half, or one, according to whether the image is recorded with no tilt removal ($\alpha=0$)[6]; with tilt removal and "far field" atmospheric propagation conditions[6] ($\alpha=\frac{1}{2}$)[6]; or with tilt removal and "near field" atmospheric propagation conditions[6] ($\alpha=1$).

In addition to the atmospherically induced blur, we assume that blur is introduced due to diffraction by the finite aperture size of the telescope. For a perfect circular telescope, we have an optical transfer function

$$\hat{B}_T(\nu) = \frac{2}{\pi}\left[\cos^{-1}\left(\frac{\nu}{\nu_o}\right) - \frac{\nu}{\nu_o}\sqrt{1 - \left(\frac{\nu}{\nu_o}\right)^2}\right] \tag{48}$$

for $\nu \le \nu_o$, zero otherwise, where $\nu_o = D/(\bar{\lambda}F)$ represents the diffraction limited cutoff frequency. For computational purposes, a convenient approximation due to Hufnagel[7] can be used,

$$\hat{B}_T(\nu) \cong 1 - 1.25\left(\frac{\nu}{\nu_o}\right) + 0.25\left(\frac{\nu}{\nu_o}\right)^4 \quad . \tag{49}$$

The total OTF of the imaging system is simply the product of Eqs. (47) and (48) or (49),

$$\hat{B}(\nu) = \hat{B}_T(\nu)\hat{B}_A(\nu) \quad . \tag{50}$$

Incorporating the definition of ν_o in Eq.(47), we obtain the total OTF,

$$\hat{B}(\nu) = \left[1 - 1.25\left(\frac{\nu}{\nu_o}\right) + 0.25\left(\frac{\nu}{\nu_o}\right)^4\right] \times \exp\left\{-3.44\left(\frac{\nu}{\nu_o}\right)^{5/3}\left(\frac{D}{r_o}\right)^{5/3}\left[1 - \alpha\left(\frac{\nu}{\nu_o}\right)^{1/3}\right]\right\} \tag{51}$$

for $\nu \le \nu_o$, zero otherwise.

We have used numerical integration to calculate $(\Delta\nu)^2$ (Eq.(33)), and Q (Eqs. (38a,b)) for the case of a point-source object ($\hat{\Phi}_o = 1$) with blur $\hat{B}$ of Eq.(51), and an ideal transfer function $\hat{S}$ of Eq.(26). In this case,

$$(\Delta\nu)^2 = 2\pi(1+\bar{N}) \int_0^{\nu_o} \frac{|\hat{B}(\nu)|^2 \nu \, d\nu}{1 + \bar{N}|\hat{B}(\nu)|^2} \tag{52}$$

and we can show that

$$Q = \frac{(\Delta\nu)^2}{\pi\nu_o^2 - (\Delta\nu)^2} \tag{53}$$

In the computations we have assumed the parameter values

r_o = 5 cm, 10 cm, 20 cm and ∞

D = 152.4 cm

$\bar{\lambda}$ = 5×10^{-5} cm.

It is not necessary to assume a specific focal length for the telescope if we work with spatial frequencies $\Omega = F\nu$ measured in cycles per radian of arc in the sky. The ratio of ν/ν_o in Eq.(51) is replaced by Ω/Ω_o, where $\Omega_o = F\nu_o$, and we calculate $(\Delta\Omega)^2 = F^2(\Delta\nu)^2$ rather than $(\Delta\nu)^2$. Likewise, Q is redefined as

$$Q = \frac{(\Delta\Omega)^2}{\pi\Omega_o^2 - (\Delta\Omega)^2} \quad , \tag{54}$$

with no change in its numerical values resulting.

In Fig. 5 we show plots of the "maximum restorable frequency" $\Delta\Omega/\sqrt{\pi}$ vs. $\bar{N}$ for the case of a point-source object and the parameter values specified above. The incident flux is varied over eight orders of magnitude. Figure 5a corresponds to the "no tilt removal" case, while 5b and 5c represent the "tilt removed" cases for far-field and near-field atmospheric propagation conditions, respectively. Note that in most cases, the maximum restorable frequency increases very slowly with $\bar{N}$, implying that, for the parameter values specified here, truly enormous amounts of light flux are necessary to record an image which can be substantially restored.

Note that the removal of tilt provides relatively little benefit in the "far-field" atmospheric propagation case. Under such conditions, which are often encountered on horizontal propagation paths, amplitude and phase fluctuations play equal roles in degrading the image, and tilt plays a relatively minor role. However, under "near field" atmospheric propagation, which is often encountered looking vertically from a good observatory site, amplitude effects are negligible and tilt removal provides some positive benefits.

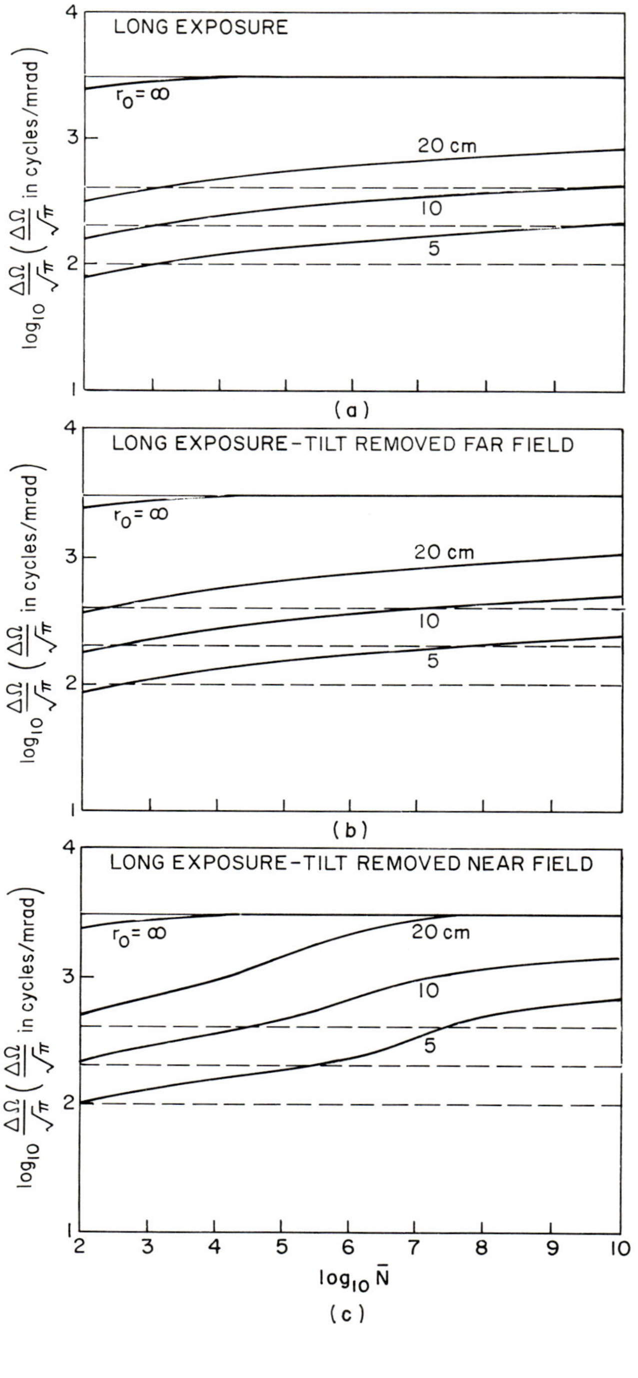

Fig. 5. Maximum restorable frequency $\Delta\Omega/\sqrt{\pi}$ vs. $\bar{N}$ for r_o = 5cm, 10cm, 20cm, and ∞, D = 152cm and $\lambda = 5\times10^{-5}$cm; (a) long exposure, no tilt removal, (b) long exposure, tilt removed, far-field atmospheric propagation, (c) long exposure, tilt removed, near-field atmospheric propagation, The solid horizontal line represents the telescope cutoff frequency D/λ, while the dashed horizontal lines represent r_o/λ.

In Fig. 6 we have plotted the quality factor Q for the same range of $\overline{N}$, again for four values of r_o. Figures 6a, b and c again correspond to no tilt removal, tilt removed with far-field atmospheric propagation, and tilt removed with near-field atmospheric propagation.

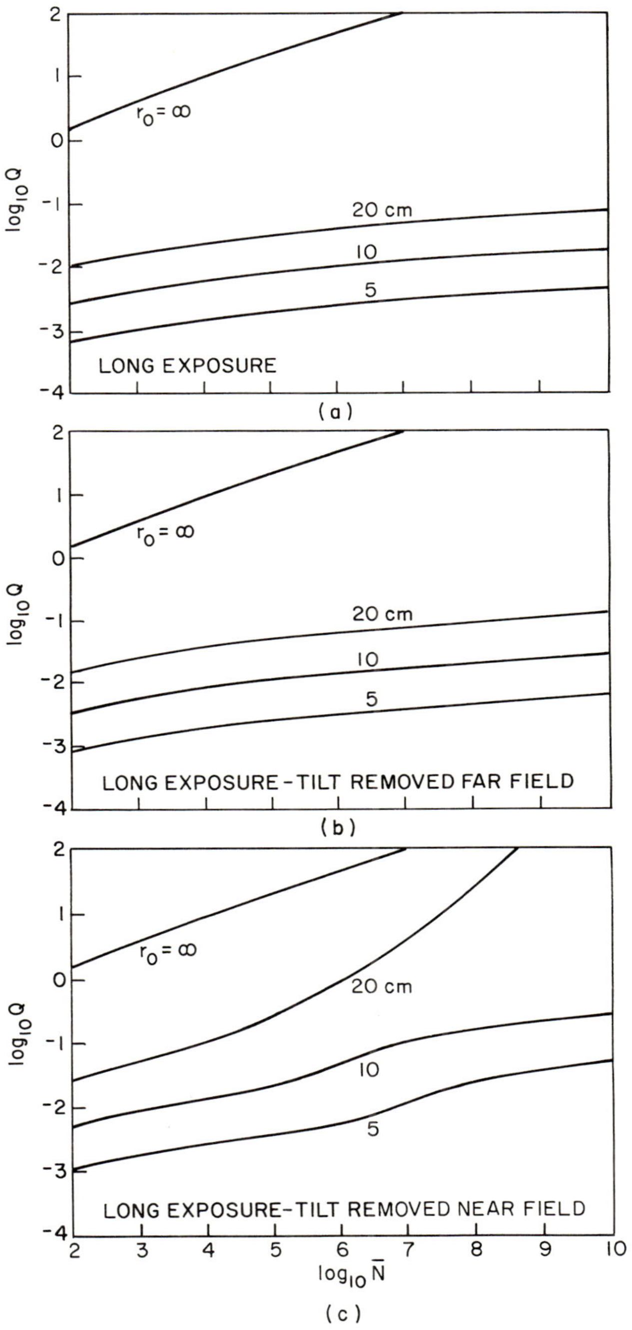

Fig. 6. Quality parameter Q, for numbers identical with those in Fig. 5 parts (a), (b) and (c) corresponding to the same cases indicated there.

As a general conclusion, we note that the results shown here support the experimentally observed fact that long-time-average atmospherically degraded images with no tilt removal are extremely difficult to restore in practice. In principle, if enough light were utilized in detection of the image, significant restoration would be possible. However, for the conditions of interest here, the theoretical results imply that the light flux required is prohibitively large. This conclusion must be modified if tilt removal is included and the atmospheric propagation conditions are what we have called "near field". Under such conditions, performance near the diffraction limit can be obtained for reasonable amounts of detected light provided that r_o is rather large (see the r_o = 20 cm curve in our results).

We note in closing that, although the results above have been derived for the case of a point-source object, they can be shown to be a close approximation for the case of a more complicated object provided $\sigma_o/\bar{o} \simeq 1$ and $\bar{N}$ is replaced by $\bar{n}$, the average number of photoevents per correlation area of the object.

Acknowledgment

This work sponsored by the Air Force Systems Command's Rome Air Development Center, Griffiss AFB, N.Y.

References

1. L. Mandel, Proc. Phys. Soc. 72, 1037 (1958).

2. L. Mandel, Proc. Phys. Soc. 74, 233 (1959).

3. W.E. Lamb, Jr., and M.O. Scully, "The Photoelectric Effect Without Photons", in Polarization, Matiere et Rayonnement (Societe Francaise de Physique, Paris) Presses Universitaires de France (1969), pp.363-369.

4. A. Papoulis, Probability, Random Variables, and Stochastic Processes, McGraw-Hill Book Co., New York, N.Y. (1965) p. 548.

5. J.J. Burke, "Estimating Objects from their Blurred and Grainy Images", Proceedings of the 1975 International Optical Computing Conference, (IEEE Catalog No. 75 CH0941-5C), Washington, D.C., April 1975, pp.48-51.

6. D. Fried, J. Opt. Soc. Am. 56, 1372 (1966).

7. R. Hufnagel, Appl. Opt. 10, 2547 (1971).

DIGITAL IMAGE PROCESSING OF SIMULATED TURBULENCE AND PHOTON NOISE DEGRADED IMAGES OF EXTENDED OBJECTS*

James R. Breedlove, Jr.
Los Alamos Scientific Laboratory
University of California
Los Alamos, NM 87545

Abstract

Conventional post detection image processing techniques have been applied to three simulations of photon noise limited images of the moon which might be obtained through atmospheric turbulence. This study demonstrates that homomorphic filtering, parametric Wiener filtering, and constrained least squares filtering can produce nearly diffraction-limited imagery if the atmosphere is "frozen" for 1/100 - 1/10 sec.

Introduction

Photographs of extended space objects such as the moon or the planets taken from earth-based telescopes are degraded by a number of phenomena. The greatest degradations arise due to the random spatial and temporal variations in the index of refraction of the atmosphere which, on the average, reduce photographic resolution to a value on the order of one arc second. Another source of degradation is the noise introduced into the image by two processes: a) fluctuations in the response of the sensor to the incident photon flux and b) fluctuations in the incident photon flux due to fundamental quantum mechanical processes. The least important degradations are due to the telescope and its imperfections, the noise-free resolution of the sensor, motion instability, and the like.

Given sufficient motivation, most of the problems in conventional space object photography could be overcome. It is recognized that at any instant the atmosphere will pass spatial frequencies far above those corresponding to the average seeing. Therefore, given a) a means of photographing with short exposures (sensitive detector) and b) a means of measuring the instantaneous point spread function (PSF) of the atmosphere one could, theoretically, proceed to use image restoration techniques to reliably improve space object imagery obtained in the conventional way. However, noise limits the success of computer image restoration. This paper will demonstrate this limitation.

There are other approaches to solving the problem of imaging through the atmosphere. These basically are techniques for restoring the image prior to recording it; they use unconventional optical elements. It is most probable that the ultimate solution to imaging through the atmosphere will rely on both pre- and post-detection processing. In either case, the image *arriving* at the compensator or the sensor will be noisy; however, this noise is usually ignored. By assuming the detector noise is limited to fluctuations in the photoelectron count, and by dealing with simulated imagery, one can obtain a good idea of the best that space object photography could ever hope to do when coupled with current post-detection image processing technology. One should note that this study and recent simulations of predetection compensation ignore noise in the incident photon flux.

In this paper no new image processing schemes will be developed. Instead, traditional sophisticated digital image processing schemes, which are derived from greatly oversimplified models of the image noise processes, will be applied to three simulated images with different relative noise levels. From this, an assessment of the limits of current post-detection processing can be made. The techniques used are the Wiener filter, homomorphic filter, the parametric Wiener filter, and the constrained least squares filter.

Simulation of Turbulence Degraded Imagery

B. L. McGlammery of the Visibility Laboratory, Scripps Institution of Oceanography, University of California, San Diego, has simulated the images of the moon that might be obtained with a ground-based telescope, which are used in this study. The simulations started with a photograph of the moon's surface taken from the Apollo 17 command module (Fig. 1).

*Work performed under the auspices of the Energy Research Development Administration under Contract W7405-ENG-36.

Fig. 1. Original photograph of moon's surface

McGlammery then generated a point spread function for a particular realization of the instantaneous wave front disturbances which might be obtained in a turbulent atmosphere having a Kolmogorov spectrum

$$\Phi(u) = C u^{-\frac{11}{3}} \tag{1}$$

where u is the spatial frequency. The atmospheric structure function was characterized by $r_o = 0.1$ meter. Details of McGlammery's work may be found in the references[1,2]. Fig. 2 is the particular polychromatic point spread function used in this study.

Fig. 2. Polychromatic point spread function

The degraded image present in the image plane of a 1.5 meter telescope was simulated by convolving the PSF with a radiance map obtained from Fig. 1 and assuming conventional solar illumination, atmospheric transmission, and lens transmission. The validity of these simulations rests on the assumption that the illumination is incoherent (as it is) and on the assumption that the PSF is actually invariant over the object and does not change during the exposures. The latter assumptions are not rigorously true for all objects at all times, but have sufficient validity to make this study a useful one.

In addition to degradations due to turbulence, the telescope itself will blur the image and, more significantly, introduce a high frequency cut-off in the photograph. Fig. 3 shows what the scene would look like if recorded with a 1.5 meter telescope in the absence of an atmosphere.

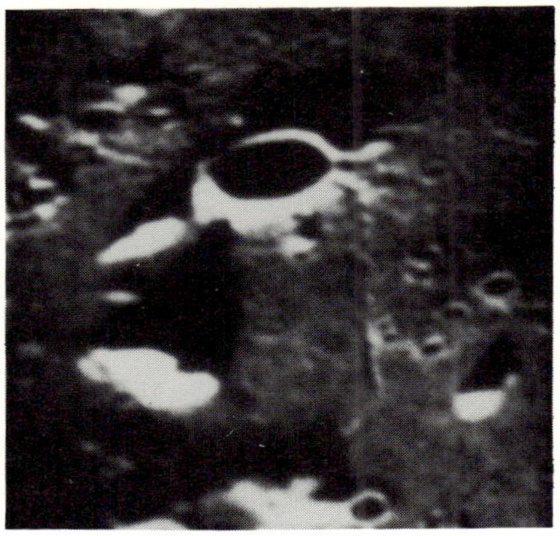

Fig. 3. Diffraction limited photograph of the moon

The detector will also degrade the image by introducing noise and by imposing its own transfer function on the scene. As mentioned previously, except for noise, detector effects will be ignored.

The formation of the simulated degraded image is described by the equation

$$g(x,y) = h(x,y) * f(x,y) + n(x,y) \tag{2}$$

where g is the recorded image, h is the point spread function of the atmosphere and telescope pupil function, f is the ideal image, and n is photon noise. The quantity n(x,y) is determined by Poisson statistics. Each pixel $(h*f)_{ij}$ is replaced by a number obtained from a Poisson distribution having mean and variance equal to $(h*f)_{ij}$.

Figs. 4a, 5a, and 6a are the simulated turbulence and noise degraded images. All three images were blurred with the same PSF, but the simulated exposure times were 1/1000 sec, 1/100 sec, and 1/10 sec, respectively. Thus, the signal to noise ratio (S/N) increases from Fig. 4 to Fig. 6. If one considers the ratio of the sum over all frequencies of the blurred image power spectrum to that of the noise

$$S/N = \frac{\Sigma|HF|^2}{\Sigma|G-HF|^2} \tag{3}$$

one finds values of 208.5 for the 1/1000 sec exposure, 2107.4 for 1/100, and 21028.2 for 1/10. However, these numbers can be misleading since the noise process has zero mean and since noise dominates at high frequencies where the data is interesting. Excluding the zero frequency term from the sum in Eq. (3) one finds $S/N' = 8.7$ for Fig. 4a, 88.8 for Fig. 5a, and 886.5 for Fig. 6a.

Digital Image Restoration Techniques

The problem of image restoration is to solve Eq. (2) for the undegraded image f(x,y). This problem is complicated not only by noise but also by the facts that a) the noise process described in Eq. (2) is not realistic, b) the point spread function is unknown in most interesting problems and is most likely ill-conditioned (has zeroes in its Fourier transform), and c) realistic models of the image recording process show that nonlinearities are introduced which make restoration much more difficult. An equation of the form

$$g(x,y) = S\left[h(x,y) * f(x,y)\right] \textcircled{OP}\, n(x,y) \tag{4}$$

is probably more realistic although as yet unsolved. In their recent work, Goodman and Belsher[3] have determined a more realistic model of noise in photon limited images; their results have not as yet been incorporated into image processing algorithms.

The difficulty in solving Eq. (2), as for all integral equations of the first kind, is that it is ill-conditioned so that even slight amounts of noise can lead to solutions which are totally obscured by noise. Satisfactory solutions to Eq. (2) depend on reducing the ill-conditioning. The filters used in restoring the turbulence degraded images are solutions to (2) which reduce the ill-conditioning in different ways.

Minimum Mean Square Error Method

A familiar image restoration technique is the classical Wiener method. This method minimizes the statistical expectation of the square of the difference of f and the estimate of f

$$\text{Minimize: } E\left[(f - \hat{f})^2\right] \tag{5}$$

where $\hat{f}$ is the restored image. Solution of Eq. (5) leads to the following frequency domain estimate of the restored image:

$$\hat{F}(u,v) = \frac{\bar{H}(u,v)\ G(u,v)}{|H(u,v)|^2 + \Phi_n(u,v)/\Phi_f(u,v)} \tag{6}$$

where the capital letters denote Fourier transforms of the quantities designated by the lower case letter in Eq. (2). The Φ's are the power spectra of the image and noise processes. The bar over a quantity denotes its complex conjugate. Comparing Eq. (2) and Eq. (6), one can write the Wiener solution as:

$$\hat{F}(u,v) = \frac{\bar{H}(u,v)\ \Phi_f(u,v)\ G(u,v)}{\Phi_g(u,v)} \tag{7}$$

It is interesting to note that Eq. (7) is identical to the result obtained by Goodman and Belsher for Poisson noise degraded images. The Wiener restoration of Eq. (7) requires one to have a great deal of knowledge about the undegraded scene (its power spectrum) and the point spread function, but the Wiener restoration remains the standard restoration against which other techniques are compared, in spite of the fact that other restoration techniques have proved superior.

Homomorphic Filter

Another image restoration filter form due to Stockham et al.[4] is commonly called the homomorphic filter.

$$\hat{F}(u,v) = \sqrt{\frac{\Phi_f(u,v)}{|H(u,v)|^2\ \Phi_g(u,v)}}\ \bar{H}(u,v)\ G(u,v) \tag{8}$$

The homomorphic filter is the geometric mean of the Wiener filter Eq. (7) and the inverse filter:

$$\hat{F}(u,v) = \frac{\bar{H}(u,v)\ G(u,v)}{|H(u,v)|^2} \tag{9}$$

Constrained Least Squares Restoration

In a technique devised by Hunt[5], the *a priori* intuition that wild local fluctuations in intensity are not likely to occur in real images is included in a solution to Eq. (2). The technique is to minimize an expression for the coarseness in the image

$$|c(x,y) * f(x,y)|^2 \tag{10}$$

in which c is some suitable derivative-like function. The minimization is made subject to the constraint

$$||g - h * \hat{f}||^2 = Ne^2 \tag{11}$$

The double bars indicate the norm and e^2 is the variance of the noise added to the image. This noise is assumed to be a constant. The solution to Eq. (10) with Eq. (11) is

$$\hat{F}(u,v) = \frac{\bar{H}(u,v)\ G(u,v)}{|H(u,v)|^2 + \gamma\ |C(u,v)|^2} \tag{12}$$

where $1/\gamma$ is a Lagrange multiplier which arises in the solution of Eq. (10) subject to the constraint of Eq. (11). The value of γ is determined iteratively. Note that the solution for $\hat{F}(u,v)$ requires no knowledge of $\Phi_f(u,v)$ and is the noise controlled restoration requiring the least *a priori* data to implement.

Parametric Wiener Filter

A technique related to both the constrained least squares approach and the Wiener filter leads to a solution of the form

$$\hat{F}(u,v) = \frac{\overline{H}(u,v)\ G(u,v)}{|H(u,v)|^2 + \gamma\ \Phi_f^{-1}(u,v)} \tag{13}$$

where γ is chosen iteratively so that Eq. (11) is satisfied. Eq. (13) is the parametric Wiener filter with the assumption that the noise power spectrum is a constant.

Examples of Image Restoration

Restorations of the simulated turbulence blurred images using the methods described in the previous section are shown in Fig. 4b-4e, 5b-5e, and 6b-6e. To make these restorations the following techniques were used:

1. Fourier transforms were computed using a two-dimensional fast Fourier transform (FFT) of the 256 x 256 pictures and point spread function. The point spread function N had nonzero values over only about 64 x 64 elements.

2. Power spectra were estimated by calculating the magnitude squared of the appropriate FFT using a Fourier window.

3. The c-matrix in Eq. (10) is the two-dimensional discrete Laplacian operator.

4. All data values were preserved without clipping and linearly stretched prior to display.

5. The image pixels were duplicated four times to produce a 512 x 512 array which was displayed on a t v monitor and photographed. Some distortion and artifacts due to the t v monitor can be seen in the images.

Discussion of Results

It is clear from comparing the processed images with Fig. 3 that resolution qualitatively approaching the diffraction limit of the telescope is possible in the highest S/N image. There is little qualitative difference in all results in Fig. 6, since in the absence of noise all filters used tend to the inverse filter of Eq. 10. The homomorphic filter seems to produce a slightly more pleasing result in Fig. 6 than the others. The constrained restorations (constrained least squares and parametric Wiener filters) perform better than the Wiener restoration at lower S/N ratios.

Since all the algorithms used for restoring the simulated turbulence and noise degraded images are developed from oversimplifications of the noise model in the imaging process, it is surprising that they do so well restoring the images. It is anticipated that current work at the Los Alamos Scientific Laboratory in adapting Goodman's more realistic models of photon limited images to constrained restoration schemes will prove that improved restorations are possible.

References

1. B. L. McGlammery, "Computer Simulation of Atmospheric Turbulence and Compensated Imaging Systems," Proc. AGARD Technical Meeting on Optical Propagation in the Atmosphere, Oct. 27-31, 1975, Lingby, Denmark.

2. Proc. OSA/SPIE Topical Meeting on Image Processing, Feb. 24-26, 1976, Asilomar, CA.

3. J. W. Goodman and M. L. Belsher, "Fundamental Limitations in Linear Invariant Restoration of Atmospherically Degraded Images," these proceedings.

4. T. G. Stockham, Jr., T. M. Cannon, R. B. Ingebrelsen, "Blind Deconvolution Through Digital Signal Processing," Proc. IEEE, 63, 678 (1975).

5. B. R. Hunt, "The Application of Constrained-Least-Squares Estimation to Image Restoration by Digital Computer," IEEE Trans. Comput., C22, 805 (1973).

Question by A. M. Schneiderman: Did you add independent Poisson noise to the point spread function?

Answer by J. R. Breedlove, Jr.: No, it's noise free.

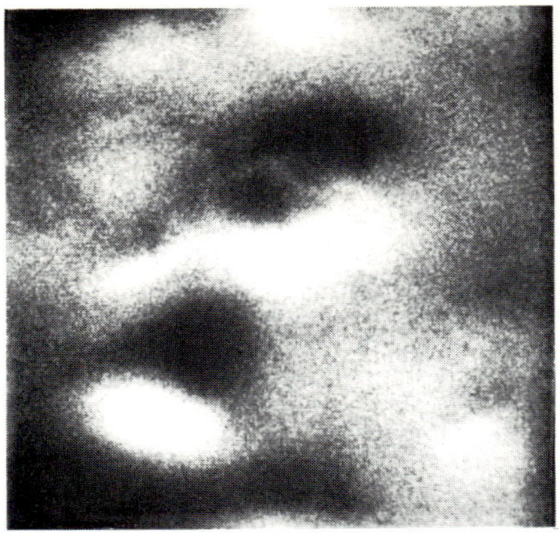

Fig. 4a. Unprocessed turbulence and noise degraded image (S/N)′ = 8.7

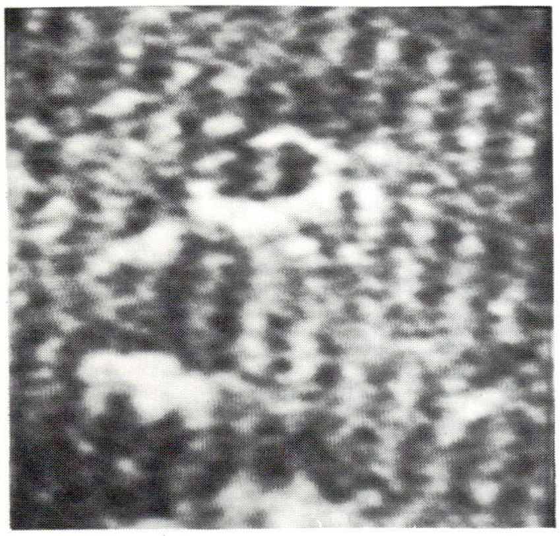

Fig. 4b. Wiener filter restoration

Fig. 4c. Homomorphic restoration

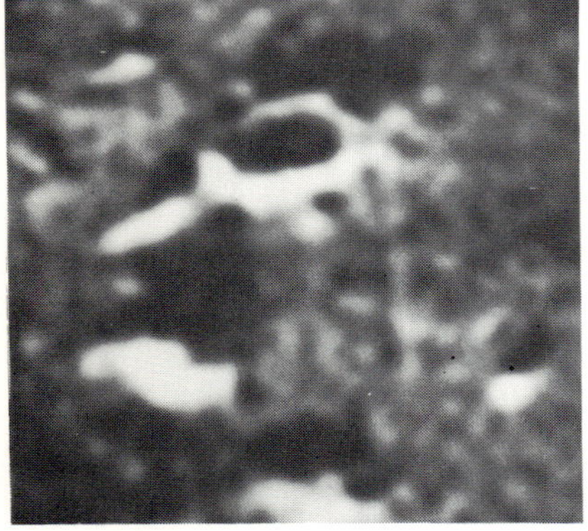

Fig. 4d. Constrained least square restoration

Fig. 4e. Parametric Wiener filter restoration

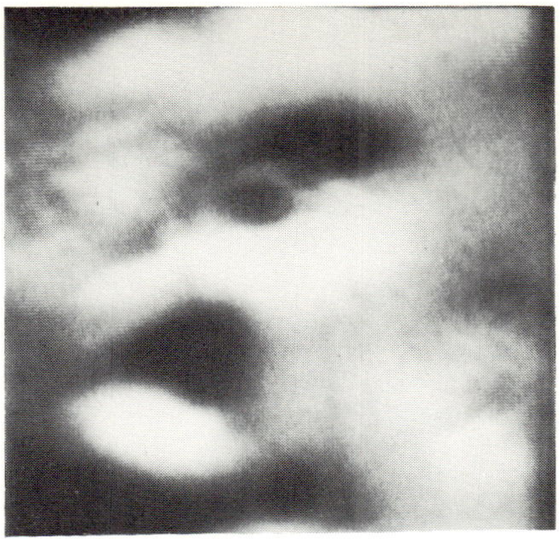

Fig. 5a. Unprocessed turbulence and noise degraded image (S/N)´ = 88.8

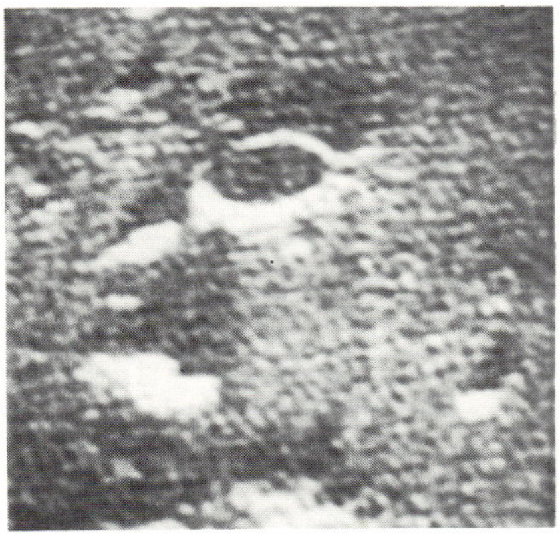

Fig. 5b. Wiener filter restoration

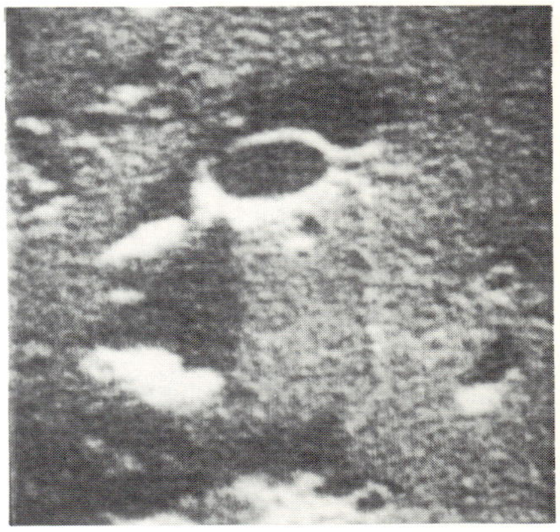

Fig. 5c. Homomorphic restoration

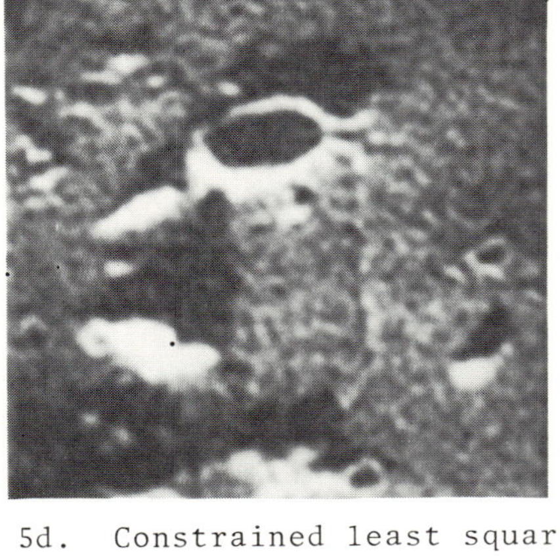

Fig. 5d. Constrained least square restoration

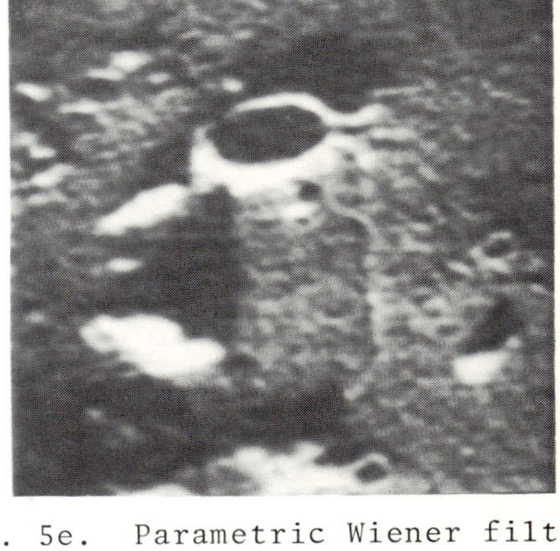

Fig. 5e. Parametric Wiener filter restoration

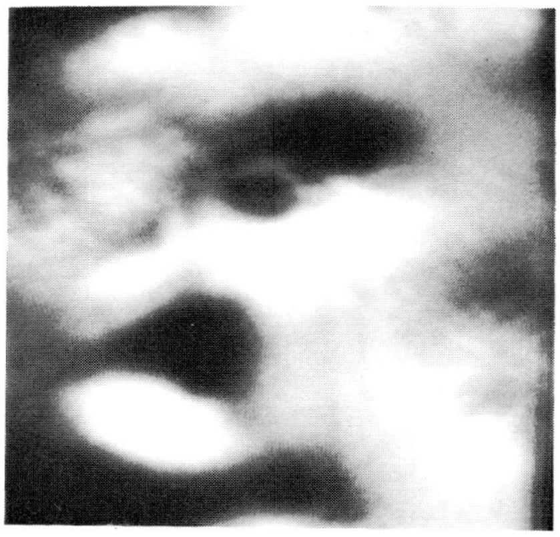

Fig. 6a. Unprocessed turbulence and noise degraded image (S/N)´ = 886.5

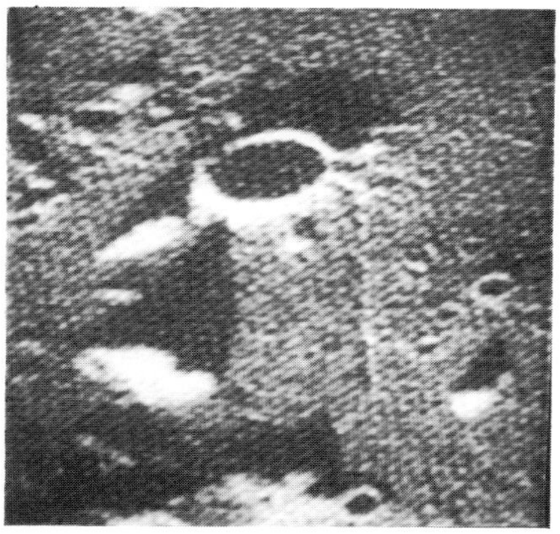

Fig. 6b. Wiener filter restoration

Fig. 6c. Homomorphic restoration

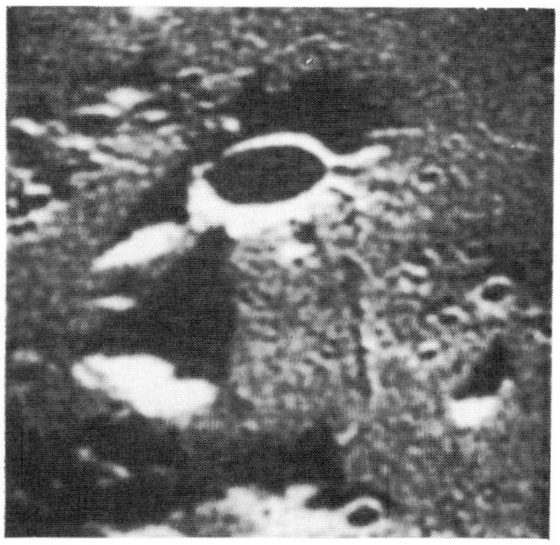

Fig. 6d. Constrained least square restoration

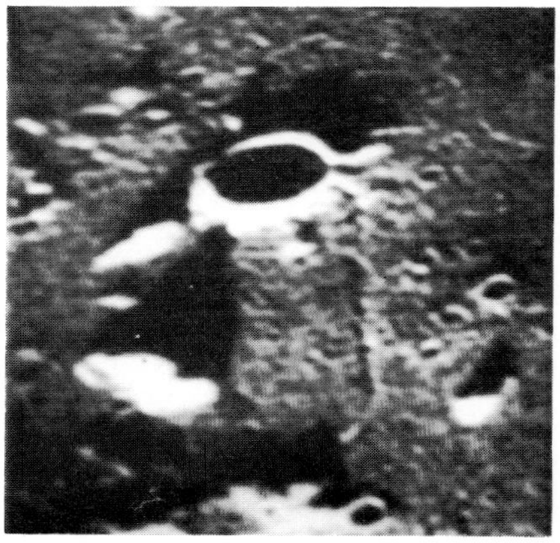

Fig. 6e. Parametric Wiener filter restoration

A STATISTICAL METHOD FOR POST-DETECTION COMPENSATION FOR ATMOSPHERIC DISTORTIONS OF IMAGES OF FAINT SCENES

Charles E. KenKnight
Lunar and Planetary Laboratory, University of Arizona
Tucson, Arizona 85721

Abstract

A new method for post-detection compensation of atmospheric distortions of images of faint scenes has been outlined and initially tested. A sequence of short exposure (0.01 to 0.1 sec) visible light images is processed in terms of the statistics of the Fourier transform amplitudes. A "master image" is derived that is iteratively compared with each image (in Fourier space) so as to align the set of images on the basis of features in the scene. Aperture synthesis can be used to decrease aperture redundancy since the alignment uses only Fourier amplitudes that are prominent in the joint set of master image and the raw image sequence. The master image has an effective point spread function (PSF) comparable to the best PSF in the sequence but the phases are strongly quieted by the statistics of large numbers if 30 or more images are spread over a time interval of 15 or more sec. Thus spatial frequencies in excess of 1 cycle per arcsec may yield reliable photometry after correction for contrast loss and telescope aberrations. The degree of enhancement may be optimized, based on a separation of signal and noise in the data so that noise may be estimated. At faint levels quantum noise is severe. Since that noise is correlated with signal, the noise spectrum is not white but falls with increasing spatial frequency.

Introduction

We shall be interested here in treating statistically the image data obtained as a sequence of short integrations in the focal plane of a large telescope. Of special interest will be scenes of low surface brightness, such as outer planets and asteroids, for which the quantum noise in the image data is not small. The goal of the image treatment shall be primarily accurate photometry in the field of view (FOV) to the highest possible spatial frequency. In view of the quantum noise problem it will be enough to improve on the mean PSF by a factor of 2 to 10. It is assumed that calibrating data on an unresolvable star will be used in an optimum fashion to assess and correct for contrast loss where it can be reliably estimated.

The wavefronts of the light from the object are multiplied by the phase screen of the atmosphere. If all the diffraction effects were within a few focal lengths of the telescope, almost all the diffracted light would be available in the focal plane. But distant phase screens are randomly apodized by the diffraction of light into and out of the telescope aperture. The details of the apodization vary with the angle through which the light rays are turned, i.e., on the spatial frequency in the telescope focal plane.

To estimate the energy diffracted by angle θ by a random phase screen filling a telescope of diameter T, consider a subdivision of the aperture into squares having sides $L=T/(2N)$ in length, where N satisfied the grating equation for angle θ:

$$(T/N)\sin\theta = \lambda \tag{1}$$

If we model the phase screen as a phase shift of $\pm z/\lambda$ when averaged over a square of side L, the Fourier transform of the transmission function collapses to a discrete sum of N phase differences along a row and there are 2N rows. Of the $2N^2$ differences, about N^2 are zero in this model. The remaining differences give a random walk about zero that is apt to fluctuate to ± N steps from the mean. Thus the electric field at θ relative to that on axis tends to be about equal to $z/(N\lambda)$ or $z/(T\theta)$ unless the optical path difference z approaches or exceeds $\lambda/2$, when saturation effects occur because $\exp(iz/\lambda)$ has at most unit modulus. In the angular range where the model does account for the gross energy distribution, $1'' \lesssim \theta \lesssim 1'$, it also predicts that the diffracted electric field is symmetric about the origin to within a factor of $z/(N\lambda)$. This also fails at $\theta \lesssim 1''$ because a path difference of λ in distance 2L approximates a facet of a blazed grating and asymmetry effects rapidly accumulate.

Though symmetry of the PSF fails for short exposure images for the two reasons above, we shall assume that symmetry is approached for a series of images separated by more than about 0.1 sec and spanning at least 15 sec.

Next we note that the random walk effects noted above have the property that if a particular spatial frequency in the Fourier transform of the intensity distribution has large amplitude, then the phase distortion tends to be small. If so, we ought to use some weighting procedure that emphasizes large amplitudes in estimating relative phases. The argument is that for highly perfect optics the contrast is decreased in an image by convolution with the PSF of finite phase perturbations. The larger the phase errors, the wider the PSF, and then lower contrast is seen to follow.

If the input data are filtered by a weighting that selects data corresponding to "better seeing", then the filtering is not linear. The present approach begins in the context of homomorphic linear filtering as discussed by Oppenheim et al.[1] Because a convolution in image space is a product of Fourier transforms, they introduce logarithms of Fourier transforms to recover a generalized superposition principle. The real part of a complex logarithm is just the logarithm of a modulus, but the imaginary part is the phase of the Fourier transform. Concerning this phase they note,

"The phase curve can be computed by first computing the phase modulo 2π and then 'unwrapping' it to satisfy the requirement that it be continuous and odd. Simple algorithms for doing this are easily generated, provided that the frequency spacing of adjacent points is sufficiently small."

They then go on to exhibit an alternative way to compute the complex logarithm in terms of the ratio of the derivative of the spectrum and the spectrum, but they note that the latter procedure is more severely affected by the aliasing associated with the discrete Fourier transform than is the former. It does, however, avoid the explicit computation of the unwrapped phase curve. The former procedure is equivalent to the phase calculation introduced for image processing by Knox and Thompson.[2] The present method is also closely allied to the former procedure. However, matrix methods replace phase "unwrapping". An alignment procedure attempts to put most Fourier vectors on the right phase branch and a mean vector is sought for each spatial frequency independently.

The Method

We shall insist on a method having features that it
(1) Produces images, not an autocorrelation function
(2) Treats each spatial frequency independently
(3) Recognizes better seeing moments impartially
(4) Uses temporal and spatial phase continuity where possible
(5) Removes centroid motion with high accuracy
(6) Affords self-consistent noise estimation.

Point (1) needs no defense and (3) through (6) are best studied by example. Point (2) is in sharp contrast to the method suggested by Knox and Thompson[2]. A scene of special interest is a finite object against a black background. The spectrum is roughly (sinx)/x in nature, which has regions of vanishingly small amplitude near the zeroes of sin x. In the presence of even small noise the Knox-Thompson method has difficulty following the phase (which has a 180° discontinuity) in the vicinity of the zeroes. And if the principal culprit in image distortion--telescope redundancy--is attacked by aperture synthesis, then finite ranges of the image spectrum are identically zero by optical arrangement. The Knox-Thompson method is limited to bright scenes and redundant telescopes because of the phase difficulty near zeroes of the spectrum. The present method bridges the zeroes by treating spatial frequencies independently and using only those portions of the spectrum that are appreciably different from zero. Such regions are found at the cost of two to three passes through the data, whereas the Knox-Thompson method can be a single-pass analysis. Their method uses only the correlations in each image; the concept of a "master image" against which all images are compared uses many more correlations and is much more sentitive.

To begin, let us suppose that a (one-dimensional) light distribution is described by the finite Fourier sum

$$f(x) = \sum (a_j \cos jkx + b_j \sin jkx),\quad k = 2\pi/L,\quad 0 \leq x < L$$
$$j = 0,1,2,\ldots,L/2. \qquad (2)$$

A translation of the origin could require calculation of the translated function by use of the shift theorem:

$$f(x+\delta) = \sum (a'_j \cos jkx + b'_j \sin jkx) \qquad (3)$$

where

$$\begin{bmatrix} a'_j \\ b'_j \end{bmatrix} = \begin{bmatrix} \cos j\theta & \sin j\theta \\ -\sin j\theta & \cos j\theta \end{bmatrix} \begin{bmatrix} a_j \\ b_j \end{bmatrix},\quad \theta = k\delta. \qquad (4)$$

The amplitudes (a,b) are components of a two-dimensional vector associated with a phase plane. For each image there is one vector in a given phase plane j. The effect of centroid motion is a rotation of successive vectors in that plane.

We seek a least-squares interpolation procedure to remove the centroid motion. We require (a) ease of calculation, (b) vanishing contribution if a weighted mean vector for that plane vanishes, and (c) no phase ambiguities more rapid than 2π per revolution. By a weighted mean vector we mean $A = \sum w_i a_i / \sum w_i$ and similarly for B. The weights might be some power of the modulus of a single vector $c_{ji} = (a^2 + b^2)_{ji}^{1/2}$ among various possibilities. There is one weighted mean vector per phase plane.

Make a temporary coordinate transformation so that the a-axis is along the mean vector for every j:

$$\begin{bmatrix} a' \\ b' \end{bmatrix}_{ji} = \frac{1}{(A^2 + B^2)^{1/2}} \begin{bmatrix} A & B \\ -B & A \end{bmatrix}_j \begin{bmatrix} a \\ b \end{bmatrix}_{ji} \quad . \qquad (5)$$

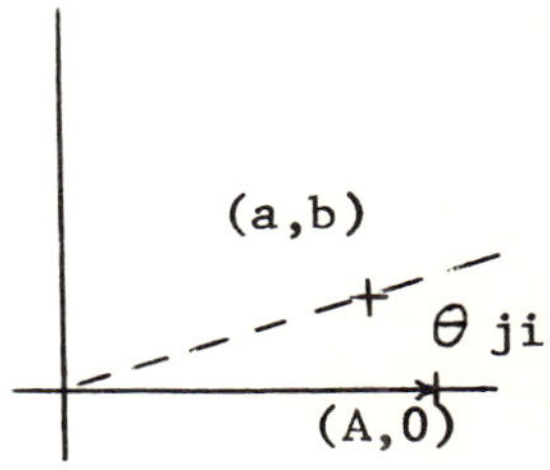

Then suppose that a certain image i is translated slightly with respect to the ensemble average of images. In every phase plane j the vector $(a,b)_{ji}$ contributed by that image will tend to be rotated with respect to $(A,0)_j$ in that plane. In a plot of the error angle θ_{ji} divided by j as a function of j (see sketch) there will be a tendency for angles of one sign. But the error in each such angle varies widely, depending on the uncertainty $(\Delta B/A)_j$, for example. Certainly no significance should be attached to the value of θ_{ji}/j associated with a vanishing A_j. But given some suitably weighted average of θ_{ji}/j, we could use the shift theorem to rotate the $(a,b)_{ji}$ contributed by that image into better alignment with the $(A,0)_j$ set. After aligning each image in this way, the mean amplitudes $(A,B)_j$ need to be recalculated and the process repeated as needed.

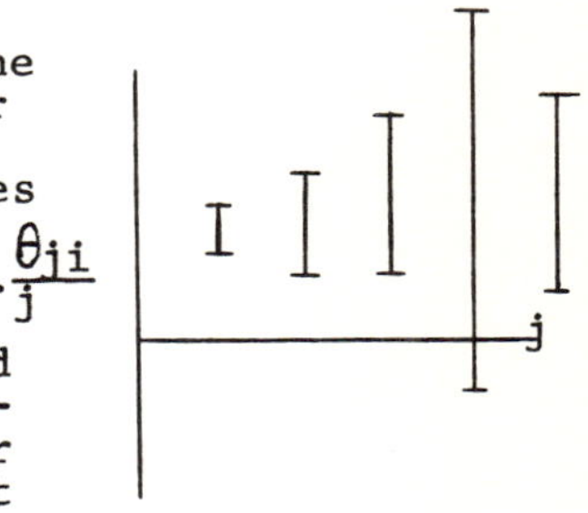

The image associated with the Fourier vectors $(A,B)_j$ is the "master image." In the first pass through the data the master image is very blurred and only the lowest spatial frequencies have $(A,B)_j$ different from zero. As registration of the images with best seeing improves, the master image sharpens and many high spatial frequencies contribute to the final registration.

The alignment procedure for image i minimizes the weighted norm $(a-A)^2 + b^2$, summing over the frequencies j. The weighting scheme generalities are (1) $A/\Delta B$ to some power must appear to suppress contribution from spatial frequencies that are not prominent in the image; (2) the single vector modulus c_{ji} to some power should appear to select the better seeing data; (3) images taken only 0.01 sec apart exhibit slow trends in the phase changes that can be removed so that several exposures can be combined coherently (critically important at the lowest light levels); (4) the weighting can be chosen so that the relatively slow calculation of square roots and arc tangents occurs never or rarely in the program (despite the indicated square roots above). The registration accuracy possible for one star-like image containing N quantum counts is about $\sigma/N^{1/2}$, where σ is the radius of the seeing disk. Given M such images, the most prominent Fourier components in a master image will have a phase uncertainty appropriate to a translation of the order of $\sigma/(MN)^{1/2}$, which may formally approach the milliarcsecond level. In practice, detector nonlinearity and nonuniformity will prevent such perfection of alignment. For general scenes there is a dc level that makes no contribution to the alignment procedure, but it does tend to spoil the procedure by contributing quantum noise. The point to be grasped about this accuracy is that centroid noise is no limiting factor until N is a few quanta per image.

Point (2) of the previous paragraph mentions weighting data by the single vector modulus c_{ji} in order to select better seeing. The effect of the atmosphere on an image is convolution with some (complicated and nonsteady) PSF, which invariably decreases image contrast. Some moments are better in that contrast is better, which means that a spatial frequency that probes the seeing disk has higher amplitude. A simple weighting scheme requiring only one pass through the data gives the weighted mean amplitudes $(A,B)_j$ from

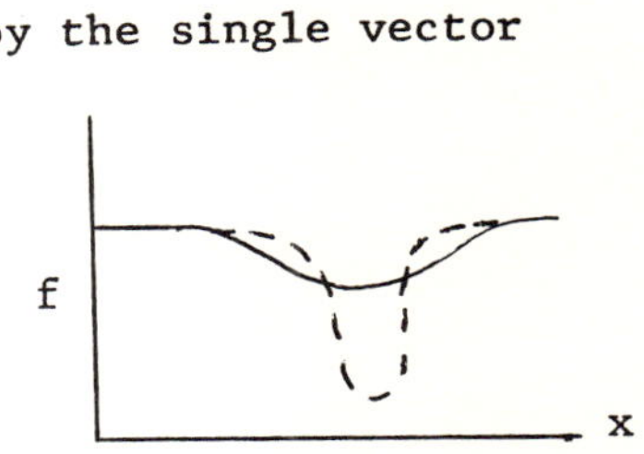

$$\left.\begin{matrix}A\\B\end{matrix}\right\}_j = \left.\frac{\sum c_i^2 \{a \text{ or } b\}}{\sum c_i^2}\right|_j \tag{6}$$

where the sums are over the vectors of spatial frequency j from each image i. Similarly the weighted power is

$$P_j = \bar{C}_j^2 = \left.\frac{\sum c_i^2 c_i^2}{\sum c_i^2}\right|_j . \tag{7}$$

The standard error in a weighted Gaussian variable y may be estimated from $s^2 = (\sum wy^2/\sum w) - Y^2$, $Y = \sum wy/\sum w$. The corresponding uncertainty in the weighted mean is not widely known: $\Delta Y \simeq s/(\sum w/W)^{\frac{1}{2}}$, where W is the maximum weight in the set of weights. If the weights w were renormalized so that W = 1, we notice that $\sum w$ is a number that expresses the effective number of measures entering the measurement set. In all cases $\sum w \leq M$, the number of images available. Equality holds only if every w = 1.

The result of weighting the components of a vector by the square of its modulus when evaluating a weighted mean component for a certain phase plane j is that a vector of less than average power is largely ignored. If modulus is the weight, then a vector of less than average modulus is ignored. Thus a mass of poor data is used only to define what "poor" is quantitatively. On the other hand, the best data are somewhat diluted by poorer data. Thus the master image will resemble the best images in the data set and will be better only in statistical stability.

The master image is defined by the Fourier vectors $(A,B)_j$, which in the 1 to 5 cycle per arcsec (cpa) range tend to have moduli much smaller than the rms moduli associated with the weighted powers, P_j. This is because the power is positive definite, A and B are not. But suppose the phase associated with an A,B were moderately secure, as indicated by the smallness of $(\Delta A^2 + \Delta B^2)^{\frac{1}{2}}$ in comparison to $(A^2 + B^2)^{\frac{1}{2}}$. The modulus C_j could be boosted to $P_j^{\frac{1}{2}}$, provided that the power associated with quantum noise Q_j were much less than P_j. Because the power is positive definite, it includes signal-plus-noise power. Properly, C_j should be boosted to $(P_j - Q_j)^{\frac{1}{2}}$ at most. At low light levels it is essential to have a means of estimating Q_j.

When the images are aligned as well as possible, a study of the placement and length of measured vectors $\vec{v} = (a,b)_{ij}$ reveals that in the directions near the weighted mean $(A,B)_j$ the vectors $\vec{v}$ are not only more numerous than in other directions, they are also longer. This correlation between modulus and "direction density" needs more study, but seems to be a straight line relationship with intercept Q_j when the distortion is severe enough so that the vectors $\vec{v}$ are distributed through 360°. A correlation of the same sort is certainly present also if the vectors $\vec{v}$ range only over a small fraction of 360°, but has barely been studied at this writing.

Beyond any doubt, the modulus variation arises from the interference associated with distortion: it has the sense that low distortion yields vectors of great length near the preferred phase. Similarly the popularity of a special azimuth expresses the familiar idea that small distortions are very much more common than very large distortions. The idea of extrapolating to zero direction density may be thought of as the limit of vanishing signal or infinite distortion. The extent that it is the "distortion" due to quantum noise or other effects that prevent following the phase is probably debatable.

Discussion

Test data have been obtained with a photon-counting Digicon image tube having a single line of 40 detectors. In various tests the detectors each subtended 1/3, 1/5, or 1/11 arcsec at a 1.5 m telescope. The accumulated counts after 9 msec were transferred to computer memory until 64 scans were accumulated. Transfer to magnetic tape required another 1.4 sec to make a cycle time of about 2 sec. The power spectra for 6.5-mag Vesta are stable but irregular using 64 scans. The PSF is not well averaged in only 0.6 sec. Eight such sets combine to give smooth power spectra because the PSF is well averaged over 15 sec.

On an occasion of high wind the centroid motion was ±2.5 arcsec but the core of light for a star was usually 1 arcsec in diameter. The centroid motion was removed; the master image for a star was also 1 arcsec in diameter.

Study of these data has just begun. One complication was a small coupling between detectors that masquerades as a variable pattern of detector sensitivities. If the relative detector sensitivities are not adequately removed, Fourier components of a special phase are enhanced so that the Fourier amplitudes rotate about a point displaced from the origin rather than about the origin, as assumed. The coupling can be removed by a matrix operation.

One dimensional data are not suitable to study the highest angular resolution obtainable by this method because the centroid also moves perpendicular to the line of detectors. Some studies urgently needed are possible. One is the effect of spectral band pass. Aperture redundancy does spoil imaging of highest resolution but it also permits imaging with white light at modest resolution and moderate zenith angles. The Airy disk is chromatic, but in a negligible fashion for most purposes, such as measuring the 1 arc-sec separation of a double star. One result so far is that the highest count rate in a star image saturates at a bandpass in excess of a few tens of nanometers. Since quantum noise is therefore saturated also but color effects are spoiling the resolution, it follows that there is an optimum bandpass for a given object and seeing conditions. There is no present survey to guide the observer in that bandpass choice.

The magnitude limit for modest inroads into the seeing disk by the present method depends upon telescope size. Signal is quite adequate for studies at 9 mag on a 1.5 m telescope with 1 to 2 arcsec seeing. Studies at 12 to 15 mag at 5 meters and with better detector quantum efficiency seem feasible.

References

1. Oppenheim, A. V., Schafer, R. W. and Stockham, T. G., Jr., "Nonlinear Filtering of Multiplied and Convolved Signals," Proc. IEEE 56:2, 1968.

2. Knox, K. T. and Thompson, B. J., "Recovery of Images from Atmospherically Degraded Short-Exposure Photographs," Astrophys. J. 193:L45, 1974.

Question by Virendra N. Mahajan:

You said non-isoplanatism was not a problem in your method. Could you elaborate why?

Answer by Charles E. KenKnight:

As in astrometry, which is a subset of this method, the phase average is based on data taken over a time interval long enough so that very many independent random phase screens enter the average. The phase perturbations of each Fourier component have opportunity to oscillate many times both positive and negative with respect to the average.

Author Index

Subject Index